GOLD

GOLD

THE RACE FOR THE WORLD'S
MOST SEDUCTIVE METAL

MATTHEW HART

**SIMON &
SCHUSTER**

London · New York · Sydney · Toronto · New Delhi

A CBS COMPANY

First published in Great Britain by Simon & Schuster UK Ltd, 2013
This paperback edition published by Simon & Schuster UK Ltd, 2014
A CBS COMPANY

Designed by Aline C. Pace

1 3 5 7 9 10 8 6 4 2

Simon & Schuster UK Ltd
1st Floor
222 Gray's Inn Road
London WC1X 8HB

www.simonandschuster.co.uk

Simon & Schuster Australia, Sydney
Simon & Schuster India, New Delhi

A CIP catalogue record for this book
is available from the British Library

ISBN: 978-1-84983-969-3
ebook ISBN: 978-1-84983-970-9

Printed and bound by CPI Group (UK) Ltd, Croydon, CR0 4YY

for Terry

CONTENTS

Not all that tempts your wand'ring eyes
And heedless hearts, is lawful prize;
Nor all that glisters, gold.

—THOMAS GRAY

GOLD

1

THE UNDERGROUND
METROPOLIS

And was Jerusalem builded here
Among these dark Satanic Mills? —William Blake

A SAFFRON LIGHT WAS SEEPING ONTO THE VELD
when I drove out to the gates of hell. The midwinter air was dry
and sharp. The silhouettes of headframes made blue shapes against
the dawn. Forty miles from Johannesburg I turned off the highway
at Mponeng mine, the deepest man-made hole on earth—a vast,
stifling oven of toiling men, thousands of them, buried miles underground as they blasted and scraped for the metal that has bewitched
and harassed man for 6,000 years. What else but gold.

Picture Manhattan Island sliced in half at the waist, and the top
part set up on its end. Mponeng mine would fill the grid between
59th Street and 110th Street, its tunnels and shafts exposed like a
giant ant colony. The half-lit streets would tower into the sky before
you, block upon block, to an altitude of almost three miles. You could

stack ten Empire State Buildings on top of each other in the distance from the bottom of the mine to the surface. This gargantuan warren devours as much electricity as a city of 400,000. Rivers of water pulse through its plumbing. Its maze of passageways howls with the noise of ventilation. Two hundred and thirty-six miles of tunnels lace the rock—thirty miles longer than the New York City subway system. Every morning 4,000 men vanish into its subterranean web of shafts, ore chutes, and haulage tunnels, and the narrow slots called stopes where miners crouch in the oppressive heat to drill the gold.

Their target was a thirty-inch-wide strip of ore. Considered against the immensity of the mine, as thin as a hair. But on the July morning in 2012 when I went down Mponeng mine, the gold price was $1,581 an ounce. That wisp of rock was spilling out $948 million worth of gold a year.

The world is awash in gold. Never has there been so much to buy and such a frenzied trade. We buy it and sell it and buy it again. In 2011 the London bullion market discovered that its members were trading as much gold every three months as had been mined in all of history. We are on the biggest gold binge ever, a scramble for gold ignited by a mouthwatering price, by the click-of-a-mouse simplicity of buying bullion, and by the sense of looming apocalypse that stormed through financial markets in the banking crisis and sent people hurrying for cover into gold.

As the gold price soared it swept up legendary billionaires. George Soros had $663 million in a single fund. John Paulson, the hedge-fund wizard, was reported to have made almost $5 billion personally on his gold bets. Fear drove the price. Banks tottered and currencies shrank, and in the three years after the collapse of Lehman Brothers the gold price gained $1,000. Spurred by the rising price, explorers ransack the planet in the greatest gold rush ever. At the bottom of Mponeng mine, in the brutal closeness, where the ore plunges steeply down into the rock below the deepest level, they are chasing it with drills. Certainly they will go down and get it.

I JOINED A SMALL GROUP that was going down. We met in a thatched reception hall. A table was spread with a breakfast of sweet pastries, little tubs of yogurt, and a steaming tray of boerewors, a pungent, coarse-cut country sausage that South Africans extol but do not always eat. A group of sturdy males from the mine's underground management stood growling among themselves and eyeing us with pleasure, as if we were Christians that they thought the lions might enjoy. Then they sat us in folding chairs to tell us about the dangers of the world we were about to enter.

Deep mining destabilizes rock. Six hundred times a month, they said, a "seismic event" will shudder through Mponeng mine. Sometimes the quakes cause rockbursts, when rock explodes into a mining cavity and mows men down with a deadly spray of jagged rock. Sometimes a tremor causes a "fall of ground"—the term for a collapse. Some of the rockbursts had been so powerful that other countries, detecting the seismic signature, had suspected South Africa of testing a nuclear bomb. "You will certainly feel a little shake when you are down there," Randel Rademann, the mine's general manager, rumbled in his gravelly, Dutch-inflected English. "A leetle shike," it sounded like. He was an enormous man in a plaid shirt, with hands like shovels and a beard that looked as if it would score glass.

After Rademann, a safety officer showed us how to use the Dräger Oxyboks self-rescuer, a device that came in a small, frail-looking aluminum case. Self-rescue was not an idea I had grappled with, and now that I was forced to, did not like. In an emergency, you were supposed to open the little case and inflate the rubber bag inside by blowing into it, then strap the bag into place like a gas mask. The device would filter out toxic gases. It would keep you alive for thirty minutes.

We changed into white overalls and steel-toed rubber boots and buckled on heavy belts. They gave us miner's hats and lamps. I attached the heavy lamp battery to my belt, threaded on the Oxyboks, and went clumping down a sort of cattle chute to the cage.

Each cage, or elevator car, holds 120 people in a three-deck stack. When the first deck has filled with its load of forty miners, the car descends ten feet and stops, and the deck above it fills in turn with another forty.

The hoisting apparatus for a cage that carries people instead of rock is called a man winder. Sometimes it winds men to their death. On May 11, 1995, in the Vaal Reefs gold mine, ninety miles south of Mponeng, an underground railway engine broke loose. Pulling rail cars after it, the engine plunged 7,000 feet down a shaft onto a two-deck cage, killing the 105 men who were returning to the surface at the end of shift. On May 1, 2008, nine miners died in a sub-shaft at Gold Fields' South Deep mine, when a winder rope broke and the cage fell 196 feet.

Once our cage was full, a dispatcher signaled with a sequence of electric bells. Slowly, very slowly at first, the cage crept down. The bright tunnel disappeared and the light faded and the black rock closed in. Almost imperceptibly the speed increased, until at last the operator, controlling the cage from a distant room in a separate building, took off the brake and dropped us down the shaft at forty-six feet a second.

My stomach sailed into my ribs. My ears blocked. Air whistled in the wire mesh. The cage squeaked and rattled as it plummeted down the shaft. Water collected on the metal frame and dripped on our heads. It was pitch dark: the only illumination came from a lamp that someone held in his hand; in its light I could see water falling like a light rain. Sometimes the light of a tunnel flashed by.

As the cage hurtled down the shaft, the suspended weight increased. The cage by itself weighed 20,000 pounds, and a full load of men roughly doubled that. Then there was the weight of the steel rope that held the cage. The rope was two and a half inches thick and weighed twelve pounds a foot. That meant that for every thousand feet we dropped, the rope added six tons to the weight it had to hold—an extra ton every 3.6 seconds. I pictured the steel rope unspooling in a blur, packing steel onto our plummeting car.

In four minutes we traveled 1.6 miles into a zone where the rock was always stirring. Immeasurable tonnages shifted, radiating spasms that miners feel as trembles in the rock. We stepped into a large, bright tunnel with whitewashed walls, and made our way to a second cage.

The depth that a single hoisting system can reach is limited by its design. In the tallest building in the world, the Burj Khalifa in Dubai, fifty-seven elevators shuttle people up and down the tower, often in stages through upper-floor "sky lobbies." We had traveled five times the distance covered by the Burj Khalifa's system, and had done it in a single drop. We made our way to the cage that would take us deeper, to the active mining levels that lay far below. We stepped into the second cage and in two minutes dropped another mile into the furnace of the rock.

In an effect known as the geothermal gradient, the temperature of the earth increases with depth. We emerged into a tunnel where the rock temperature was 140° Fahrenheit and the humidity 95 percent. In this dim sauna, perspiration soaked us in a moment. Sweat ran down my body until the socks were squishing in my boots and the cotton overalls were pasted to my skin. We had gone as deep as we could by cage—2.3 miles. To reach the deeper levels the miners either rode in open vehicles or, more often, walked. From the surface to the deepest tunnels, the journey took an hour, and most of that time was spent on the last half mile. I clambered into the back of a Toyota truck. We started down a stifling, gloomy, long decline.

In the city of the underground, there are no vistas. I peered ahead through the cab to see where we were going. All I could make out was the endless ramp, the wall rock, thick cables bolted to the granite, looping on ahead until they disappeared where the headlights lost themselves in the dark air.

The tunnel roof scraped by above our heads, lit by bulbs spaced far apart. When at last we came to a corner and turned, I felt a momentary surge of relief to be escaping the airless tunnel; but of course there was no escape. Every turn took us further into the rock, further from the cage, closer to the reef and the deep galleries. We passed a cavernous recess, where men gathered around a large tracked vehicle. They watched us go by. We turned a bend into another hot, dark tube with light bulbs dangling from the roof and cables stitched to the wall, and continued down.

AngloGold Ashanti, the mine's owner, selects miners for the deeps only after screening for susceptibility to heat. In a special chamber, applicants perform step exercises while technicians monitor them. The test chamber is kept at a "wet" temperature of eighty-two degrees. The high humidity makes it feel like ninety-six. "We are trying to force the body's thermoregulatory system to kick in," said Zahan Eloff, an occupational health physician. "If your body cools itself efficiently, you are safe to go underground for a fourteen-day trial, and if that goes well, cleared to work."

I wondered how well any screening could predict the ability to work in that asphyxiating dungeon. Nothing lived naturally in the depths of the mine except a kind of bacteria, an organism that survives without photosynthesizing sunlight. The bacteria take their energy from ambient radioactivity.

It takes 6,000 tons of ice a day to keep Mponeng's deepest levels at a bearable eighty-three degrees. They make the ice in a surface plant, then mix it with salt to create a slush that can be pumped down to underground reservoirs. There, giant fans pass air over the coolant and push the chilled air further down, into the mining tunnels. Cool air goes down at a temperature of thirty-seven and comes back, heated up by the rock, at eighty-six. I walked past one of these hot air returns—a black, growling tunnel that exhaled rank air from the bottom levels.

A deep mine is a truce that will always break. Mining at depth makes rock unstable. Every day at Mponeng mine they detonate 5,000 pounds of explosives. Every day they take away 6,400 tons of rock. The laws of compressive force dictate that the rock will try to close the spaces left by mining. To prevent this, engineers backfill stopes with rock and concrete. They reduce rock stress at the mining face, "softening" the rock before they blast it by drilling complex patterns around the blasting holes. In one deep mine they "fool the rock" by drilling out six-foot horizontal slots above the stopes. Since stress propagates through rock, but not through space, the empty spaces hinder the transmission of stress.

In tunnels, yard-long rock bolts anchor the unstable rock on the tunnel roof to the more stable interior of the rock mass. Patterns of rock bolts inserted in clusters are said to "knit" the rock together. Wire mesh and sprayed concrete stabilize the tunnel walls. Seismic sensors in the mine detect tremors at the first twitch, warning men to leave the rock face. But in the gold mines of South Africa there is a destabilizing force beyond the reach of engineers. It seems beyond the reach of anyone—a huge, pervasive, violent, and desperate invading army of thieves.

Swarming the gold mines, a skilled rabble of impoverished men and women siphon off hundreds of millions of dollars a year worth of ore. Abetted by criminal gangs, they occupy vacant mining tunnels, sometimes inside working gold mines. Because South Africa's leading mines have elaborate security, invaders can't move in and out easily. Once they penetrate a mine, they may stay down for months. Deprived of sunlight, their skin turns gray. The wives and prostitutes who live with them turn gray. In South Africa they call them ghost miners. They inhabit an underground metropolis that in some goldfields can extend for forty miles, a suffocating labyrinth in which the only glitter is the dream of gold.

HOW DIFFERENT THAT DREAM IS from what it was. Gold once had a sacred aura, like the anointed kings who wore it. The skull of the emperor Charlemagne is encased in a gold reliquary in the cathedral at Aachen that he founded in the eighth century. A gold cross tops St. Edward's Crown in the Tower of London. The cross showed that the king ruled by divine right. Gold was the metal that glorified God. Seville cathedral's golden altarpiece, sixty-five feet high and sixty wide, tells the life of Christ in twenty-eight panels that took more than eighty years to make.

The sacred power has morphed into a different kind of power, the chaotic power of a price that changes by the second, a cipher of nothing but wealth. Gold reflects the society that uses it. In early times it stood for an order that concentrated power in king and church—

those who monopolized violence and sacred authority. In our day, power is concentrated in the hands of a commercial elite, and gold stands for that commercial power. In August 2011 the "BlackBerry riots," named for the handheld devices that helped the rioters meet, and dodge police, sent tens of thousands into the streets of London to trash stores. One of the rioters' targets was the banking industry, taken as a symbol of a corrupt system. As the rioters were smashing ATMs the gold price was streaking to new heights. In South Africa, where gold is the very stuff that built the state, all the world's ills seemed to meet at the gold mine door—corruption, organized crime, violence, poverty, despair.

Surely an apocalyptic contest could be seen. On the one hand, people that history did not love: despised refugees from the poverty and war of neighboring countries; former miners discarded by a shrinking industry; masses of the wretched from the slums. And on the other hand, the owners of a substance bid up to fantastic prices by people on the run from an economic disaster some of them had helped create. Where such antagonists contend, guess who dies.

IN JOHANNESBURG I MET A killer called Bad Brad. He'd been charged with murdering four miners, but was not convicted. He was thirty-seven, about the size of a phone booth, and had dark blond hair and flat green eyes. We shook hands and I followed him to a parking lot where we made our way to a Jeep Grand Cherokee finished in a textured matte black vinyl wrap that gave it a satisfying, lethal appearance. In the back were Kevlar vests and a clear plastic bag of .223 rifle ammo. Bad Brad put on aviator shades and we drove off through the Sunday traffic to have a look at the place where the four gold miners had died, two others had been injured, and the rest had fled for their lives into the pitch black galleries.

Brad Wood was not a criminal, but a gunslinger, a man hired at various times to bring order to the Wild West town that is much of South Africa. The order he brought had a price in bodies. By his own count Bad Brad had shot to death forty-two people in his sixteen years as a hired gun. The ability to do what he did, and I suppose his reputation for doing it, made him useful to the powerful people who ran a gold mine.

We took the N12 motorway west to the town of Springs. Bad Brad showed me the extras that he had on board—a siren and flashing lights. The black Jeep was registered as an emergency response vehicle, and entitled to travel at 1.5 times the speed limit. We sailed along at this refreshing pace, and I asked Wood how he'd gotten to be a killer.

In 1995 at the age of twenty-one he had joined a special unit of the South African police and trained for work in the Durban taxi wars. Durban, on the Indian Ocean, has a population of 3.5 million. It is South Africa's main port and third largest city. As in the country's other cities, private taxis form an important transportation system from the outlying townships to the city center. Because the taxi routes are lucrative, businessmen compete for them, sometimes by killing a competitor's driver. Brad provided security by following taxis and shooting the attackers.

In January 1997, when his best friend got killed on the job, Brad quit the police and went freelance. His first client was Mandla Gcaba, a nephew of Jacob Zuma. Zuma is now the president of South Africa, but was then deputy leader of the ruling party, the African National Congress. Although it would be twelve years before he took the highest office in the land, Jacob Zuma was still a powerful figure, and his younger relative a well-connected businessman. "Zuma's nephew approached me and asked me to come and be his frontline body protection," said Bad Brad. "Someone had shot him with an AK. He's got a big ugly scar on his neck. The bullet went in his back and came out his neck. He was the president of all the long-range taxis and I was his bodyguard."

Brad was not known as Bad Brad then, nor did he earn the name in the way you might think, by shooting people. He got it by taking part in the 2001 South African version of the reality TV show *Big Brother*, a show that puts good-looking young people together in a confined space until they discover, as they do, that they are all awful. "I lasted six weeks," said Brad. "I swore very bad, I tried to break out, I smashed the cameras. I had a bit of fun. When I was finished I left, and that was that."

That was that except now he was famous. The newspapers had started calling him Bad Brad, and the name stuck. The notoriety did not hurt his business, nor sever his connections to the elite. When the new owners of the Aurora gold mine found their property crawling with illegal miners, they hired him to head security.

Brad's new employers were importantly connected. Jacob Zuma had been president of South Africa for a year. Brad had been bodyguard to one of his nephews in the taxi wars, and now, at the gold mine, he worked for another Zuma nephew, and for a grandson of Nelson Mandela. The second Zuma nephew, Khulubuse Zuma, and the Mandela grandson, Zondwa Mandela, were among the owners of Aurora Empowerment Systems, the company that had bought the Aurora mine from its previous owners, a firm in liquidation. In South Africa the term "empowerment" refers to the transfer of shareholder equity, often in mining companies, to black people, as part of a scheme to remedy the injustices of apartheid.

• ● •

THE TOWN OF SPRINGS LIES on the Witwatersrand, the richest goldfield in history, a 300-mile crescent that arcs around Johannesburg in a wide belt of gold deposits. Forty percent of all the gold ever mined in the world has come from that single geological formation. Much of the Witwatersrand, except for deep mines like Mponeng, has been mined out. The Aurora mine was a case in point. The property covered 81,000 acres in three mine licenses. In its heyday, 150 shafts had opened into hundreds of miles of tunnels that tapped the fabulous "black reef," the carbonaceous, coal-like ore of the deposit. By the time Aurora Empowerment bought the site, only eight shafts of the original 150 remained. A mill processed what production there was and a shabby office building housed the administration. Yet underground, the Aurora mine was a different story. The tunnel infrastructure of the mine remained intact. Invisible from the surface, a city of shafts and tunnels honeycombed the reef. What's more, the ground was loaded with gold. Everywhere in that sprawling catacomb, nooks and crannies contained gold-bearing material. In mining, "ore" means rock that can be mined profitably. In the course of mining, the past owners had left behind rock not rich enough to be considered ore. But now, with the gold price smashing records, it was, and the new owners of Aurora wanted it. They asked Bad Brad to escort a team underground to assess the damage caused by thieves and illegal miners.

The team met at a shaft in Springs at 9:30 A.M. on August 9, 2010, a Monday. It was a public holiday. There were five of them, including Wood; Herbie Trouw, a tall, thin, chain-smoking mine manager, who carried in his head a map of the underground warren they were about to enter; Willie Coetzer, a "captain," or foreman—a chunky, bluff, gray-haired master miner; and two security guards. Coetzer had a .38 Special handgun, "just for my own protection." Wood had his Glock and a Dashprod .223, a compact semiautomatic rifle made by a Johannesburg gunsmith and designed for "close contact." The only other weapon was a JPX pepper gun carried by one of the guards.

There was no working cage. The party made a slow descent by steel ladder fixed to the side of the shaft. They climbed down 300 feet to the first mine level. Trouw and Coetzer were the only ones familiar with the inside of a mine. "The others were security, and [it was] their first day ever to go underground," said Trouw, "so they were a bit nervous."

A couple of miles to the northeast, illegal miners were also going down. They used their own, hand-dug shaft. Both shafts led into the same maze. Two opposing forces were gathering in the mine. Although separated by miles of twisting tunnels, they shared an interest that would draw them together: some of the richest ore in the world.

One way to keep tunnels from collapsing is to leave in place stabilizing areas of unmined rock known as pillars. At Aurora, the pillars were ten feet thick and six feet high. Each pillar contained fifty tons of ore. Three of the pillars were very rich, grading about 6.5 ounces of gold per ton. In August of 2010 the gold price averaged $1,230 an ounce. Those three pillars alone contained more than $1.2 million worth of gold.

Mining the pillars would not necessarily cause the tunnel to collapse. It would create a danger that desperate men would accept. If they did not die, they would be rich.

"That zone was a high-grade channel that we mined a couple of years back," Trouw told me, "a bit like an old riverbed where the gold was deposited many years back. It was approximately thirty yards wide and maybe 200 yards long by six feet high. A good estimate would be that we left about 100 kilograms of gold behind in this area in the pillars and blasted material that was not cleaned properly, plus we left a lot behind due to faulting in the rock structures."

More than 3,000 ounces of gold, then, worth about $5 million, lay in that one gallery of the dilapidated mine. After inspecting electrical substations near the shaft, where thieves had stripped out all the copper wiring, the party struck off in the direction of the richest ore.

It lay in tunneling beneath an abandoned pit. As was usual with a deposit near the surface, it had first been mined by open pit. When the original miners had reached the deepest practical level for a pit, and were still getting ore, the mining had moved underground. When illegal miners came to exploit the deposit, they picked the floor of the pit as the nearest point to the tunnels below, and sank their shaft. In South Africa the Zulu slang for an illegal miner is zama-zama, which means "try your luck." But luck was not the main ingredient in the illegals' success. Many were experienced miners who had lost their jobs, or men led by such veterans. They knew where the rich ore was, or soon found out. Since it was the richest ore that the new owners of Aurora wanted to assess, the two groups underground that day were fated to meet.

Wood's party had come underground at 9:30. They spent about an hour in the tunnels near the shaft. At 10:30 they started for the area where the zama-zamas were working. They walked for two and a half hours and covered as many miles. It was slow, hot work. The floor of the passage was uneven; the tunnels ramped up and down; the natural heat of the rock made the air stifling.

Because they knew they could meet illegal miners, who might be armed, Trouw and Coetzer wore jeans and T-shirts rather than the usual miners' overalls, which are sewn with reflective tape that would offer a good target for someone with a flashlight and a gun. Bad Brad and the security guards wore plain blue overalls. At about 1:00 P.M. the two groups met.

"We came to a place where the tunnel went up a hill," Wood told me, "and I walked to the top and when I got there I saw that the whole place ahead of me was full of lights. I turned to get a better footing and the ground gave way, and I slid down towards them making a lot of noise, and as I fell I saw the flash of a gun and then heard it, and I thought—stuff it!—and I shot back."

In the exchange of fire, Wood said, he screamed at his party to get out. They fled back along the tunnel. Behind them, at least four intruders lay dead and two were wounded. One account said that thirteen miners might have died in the gunfire. Wood and the others were each charged with four counts of murder and five of attempted murder. In a two-week trial, the prosecution tried to prove that the Aurora team had entered the mine intending to evict the zama-zamas, and that Wood had gone into the confrontation guns blazing. This theory failed to convince the judge. He accepted that Wood was the only one in his party to fire, that he had fired between seven and thirty rounds, and that he had fired in self-defense. No one knows how many illegal miners were underground that day, but when police raided the Aurora mine four months later they arrested more than 200. At an earlier raid at a gold mine in Free State province, illegal miners attacked the police with stones. That day the police killed one and arrested 426.

On the day Bad Brad and I drove out to the Aurora mine, two dejected guards watched the gate. Ten feet from their post, the chain link fence lay on the ground where someone had driven through it. The brick buildings were a shambles of broken windows and missing doors and holes through the walls. A bedraggled population haunted the rows of roofless miners' hostels. The roofs had been stolen. Men trenched for gold in plain sight. They collected the dirt in wheelbarrows and took it to a central location to pan it in a sluice, recovering whatever residue of gold there was. A battered yellow Dodge packed with young men tore past as we drove out. Wood said they were runners, picking up the gold concentrate. A few miles away, we climbed a bank and looked down into the pit. In broad daylight, a gang was stealing ore.

The pit was the gateway to the zone of high-grade ore beneath, and a collection point for the people stealing it. Beneath the pit floor were the Aurora tunnels. We watched as miners crawled from a hole. Bags of ore lay in a row, and porters staggered up a track to a row of vans.

IN SOUTH AFRICA THE THEFT of ore from gold mines is a legendary enterprise. Thieves steal about a billion dollars a year worth of ore, or maybe twice as much. The distance between these estimates shows how little grasp authorities have of the problem. Gold mine theft is a staple topic in South Africa, and many people think the police connive at it. A few days after my visit to Springs, I boarded the high-speed intercity Gautrain at the Rosebank station in Johannesburg, and went swooshing off to Pretoria, the administrative capital of South Africa, to meet a policeman who'd agreed to talk about gold mine crime and its connection to the powerful.

At suburban Hatfield, a police van picked me up for the ten-minute drive to a sprawling, fenced campus of single-story brick buildings—the headquarters of the Hawks, the elite unit of the South African police. We drove past the guardhouse and through the complex and pulled up at a little bungalow that stood apart. It had a two-car garage and a pair of identical white Volkswagen GTIs gleaming in the driveway. With its flagstone path and crew-cut hedge it might have been a starter house in Palo Alto. We went inside and sat down in a little boardroom. The winter sun poured through slatted blinds.

Colonel McIntosh Polela, the official spokesman for the Hawks, was a trim, stylish figure. He wore a gray-striped shirt with the two top buttons open. His trousers had a knife-edge crease and his shoes glowed with a high polish. His head was shaved. A TV reporter before he joined the Hawks as their public face, he maintained a grave demeanor. He opened a thick hardcover diary whose pages were crowded to the edges in a dense, neat handwriting. Reading from these notes, he reviewed the epidemic of violence and larceny that is illegal mining in South Africa.

Driven by the high gold price and the desperation of the poor, illegal miners had overrun abandoned gold mines and vacant parts of working mines, and produced a flow of illicit gold. For South Africa the problem was not only theft, but social disruption. Many of the illegal miners, Polela said, were also illegal immigrants. They came from Zimbabwe, Lesotho, Mozambique, Malawi, and Namibia. Their presence stirred up anger in the South African poor, themselves in a condition of permanent want. Vigilantes took bloody measures against the migrants.

Polela said that another problem facing police was that they could not pursue illegal miners underground. Not only did they have no training to work in mines, but police insurers would not cover them if they did. "In the shooting at Aurora," Polela said, "if it had been policemen who'd been shot, they'd not have been covered." Because police could not pursue criminals underground except at a risk borne entirely by the officers themselves, illegal miners had been able to create no-go areas both below and above ground, enforced by violence. Nor was it a question of forty or fifty miners, as at Aurora. At Welkom, in Free State province, more than 1,500 illegal gold miners were working on the surface and 270 underground. "In a functioning mine," Polela said, "they create their own pockets, sometimes with the help of legal miners."

They had been doing so for decades, robbing the mines from within. A 2001 monograph produced for the mining industry by South Africa's Institute for Security Studies, said that well-provisioned groups of up to twenty-five miners at a time infiltrated mines by bribing the security. The groups did their own blasting underground with stolen explosives. They built small refineries, establishing themselves in vacant tunnels. They had detailed maps of the mines, and could navigate the underground maze with confidence.

Illegal miners sometimes threatened and even attacked mine inspectors who discovered them. In the echoey tunnels, thieves could easily hear the approach of mine security teams, and be ready to ambush them. This made the security teams nervous about going down. They could not know how many illegals they would meet. Gold diggers often entered the underground from one mine and moved through tunnels that connected to a second or third mine. The gold-bearing material that they mined was stored in disused shafts and then retrieved over a period of time, the paper said, "often in collusion with runners who are employed in the mine. It is then transported to one of the more than 170 smelting houses [operated inside the miners' hostels] and elsewhere in the Free State goldfields."

The writer concluded that syndicates of thieves, miners, and smugglers were stealing thirty-five tons of gold a year. At a mean price of $350 an ounce for the years he studied, 1994–1998, $400 million worth of stolen gold was flowing out of the country every year.

Legal miners made good money by colluding with the thieves. According to one newspaper report, a smuggled 48-cent loaf of bread could fetch $12 underground. And there were others earning money from the zama-zamas too. "They [the miners] have contacts with the police," Colonel Polela told me. "We have to admit that people who should be fighting the illegal mining are in fact participating in it."

With this assertion, Polela voiced a common belief of South Africans—that corruption infects the government. The police were helping to rob the mines, and behind the police, the forces that controlled them. Polela saw a nexus of the rich and powerful, of politicians and of criminals, jointly preying on the mines. The stolen gold passed through the hands of organized crime syndicates in South Africa to clients in the Persian Gulf, India, and Russia, where the gold was laundered through sham companies and sold into the legitimate bullion trade. According to Polela the theft from mines was threatening the stability of South Africa by crushing investor confidence. Mining security costs had "skyrocketed," he said, creating a situation so harmful to the interests of gold mining companies that South Africa was looking to the United Nations for help in fighting the international trade in stolen gold.

But which South Africa sought this help? Not the robbers. Polela was not the only expert to suspect the involvement of those he called "big players in South African society." Peter Gastrow, who wrote the 2001 study for the industry, reached the same conclusion: that an untouchable elite was behind the theft of gold.

When I met him in New York City in 2012, Gastrow, a South African lawyer, was a senior fellow at the International Peace Institute, and its program director. His focus was transnational crime. He had been Cape Town director of the Institute for Security Studies, a government prosecutor in South Africa, and an expert in organized crime. "My impression," he told me, "is that [the pursuit of gold thieves in South Africa] has become far more politically restricted. Imagine you are an ordinary cop. These are powerful people involved. It's an intimidating environment, not just because of the scoundrels, but in addition there are these very powerful people."

Gastrow said he thought he had detected only "the tip of the iceberg" in his investigations. "I had the impression that I was scratching the surface. Mining houses felt the same way. I left my job there feeling that [theft from mines] was going to be a huge industry in the future. Powerful people were involved. I spoke to mining houses, and they were seriously worried."

They had good reason to be. At the Barberton mine in Mpumalanga province, for example, there were once as many people stealing gold inside the mine as there were legal miners working for the owner. To break this stranglehold, Pan African Resources, the owner, increased its security budget 237 percent in one year, to half a million dollars a month. A helicopter flew continuous patrol. In 2010 the police arrested the heads of six of the seven syndicates believed to have been robbing Barberton, and the company declared victory over the thieves. It was a doubtful claim. Even policemen specializing in gold theft, Gastrow had reported, could not agree on the number of syndicates plundering the mines.

One could reach one of two conclusions about at least some of the police: either that they knew exactly how much gold was being stolen because they were helping steal it, or that they didn't know how much gold was being stolen *even though* they were helping steal it. I tick the second postulate. The truth is, nobody knows, except that it's a lot.

Willie Jacobz, a senior vice-president at Gold Fields, one of South Africa's biggest miners, told me that the industry was losing around 10 percent of its ore to thieves. Let's put this another way. At just one of its mines, South Deep, Gold Fields produced 273,000 ounces in 2011. At a conservative mean price of $1,600, that would be $436,800,000 worth of gold. By Jacobz's estimate, then, thieves were taking more than $43 million of his company's money out from under their noses. This said a lot about how crazy the gold price was. How many companies can see their profits rise 27 percent in a single quarter, as Gold Fields did in the fourth quarter of 2011, while having 10 percent scooped straight out of revenue? But hang on—it gets even more fantastic. A mining company director with intimate knowledge of the South African gold scene told me off the record that the theft was likely higher—in fact, *twice as high*. A security consultant hired by his board had reported that they were losing as much as 20 percent from one of their most advanced gold mines. By this math, more than $80 million worth of ore could be hijacked out of a mine like South Deep in a year.

I asked several mining-stock analysts if they had a clear picture of the theft. They told me that the miners do not state such losses openly, but give "post-theft" data to the analysts. Nevertheless, that the theft "is bigger now than ever before in the country's history is a given," said Leon Esterhuizen, head of the London-based mines and metals unit at CIBC bank, and a veteran analyst of South African gold mines. Esterhuizen saw an industry robbed at will by "rich criminal masterminds sitting back and offering key people amounts of money several times their salaries to just 'look the other way.'" Like Polela and Gastrow, he believed the thieves had "deep connections within the police force and politics."

In South Africa, a kleptocracy was feeding at a golden trough.

Gold had helped build the country and now was helping to corrupt it. A metal had seduced a state.

GOLD BOOSTERS LIKE TO SAY that men have always thought that gold was valuable: that its value derives from universal certitude. But we don't know what people always thought. The first gold miners were preliterate and left no record of their thoughts. It's fair to suppose they liked gold, because they kept it.

I have a little catalogue called *Thracian Treasures*, from the Varna Museum in Bulgaria. In the ancient world, Thrace was renowned for its gold mines. Some of the objects pictured in the catalogue are 6,000 years old. They form the earliest treasure of gold artifacts in the world, discovered by accident in 1972 when a backhoe operator uncovered a late Neolithic tomb at Varna on the Black Sea, unearthing bracelets and beads; a solid-gold nail with a mangled shaft; a tiny breastplate, four inches square; a delicate, paper-thin spiral tape, like a golden ribbon. The tape weighed a tenth of an ounce. What can it possibly have been good for but delight?

The Thracian objects had no practical use. They were only pretty. We can imagine what attracted those who made them—the brightness of the metal, the ease of shaping it, its resistance to corrosion. In a mortal world, it was eternally bright and beautiful. Today our asset menu is immeasurably longer than the Thracians' was, yet gold is still high on it, locked into place by a revolutionary act in Lydia in Asia Minor in 635 BC—the invention of gold money.

Lydian coinage spread through the whole Mediterranean world. The effect on the place of gold in public life was profound. For states, gold became a necessity. Yet by the fourteenth century, 2,000 years later, the entire world supply would have fit into a six-foot cube. Countries were famished for gold; in Europe, mints were closing. The financial historian Peter Bernstein called this period "the sacred thirst" for gold. It was a thirst that powered the first gold rush—a murderous, cruel, intoxicating, brutal adventure that swallowed an entire civilization and spat it out as coins.

2

RIVER OF GOLD

*So began one of the strangest motions of enemy forces
in history. On one side a god-king with 80,000 battle-
hardened troops; on the other, a handful of aliens, some of
them sick, thousands of miles from any support.*

SPANIARDS CAME WELL EQUIPPED FOR THE LARCENY
of the sixteenth century. They reduced two empires, almost with a
blow. They had the cavalier's weapon of mass destruction—Toledo
steel. The swords were strong and flexible and the blades could take
a razor edge. One good stroke took off a head. A horse and rider in
full armor weighed three quarters of a ton. This massive equipage
thundered along at twenty miles an hour, concentrating the whole
weight on a sharpened steel point at the tip of a ten-foot lance. The
Spanish could project such power through advanced technologies in
sailing and navigation. And they had a pretext for the conquests they
would make: winning souls for God. When he set out, Christopher
Columbus wrote his royal backers that he would accomplish "the
conversion to our holy faith of a great number of peoples." But he

did not forget to mention gold. He mentioned it 114 times, versus twenty-six for God.

Hernán Cortés arrived on the Yucatan coast early in 1519, and on Holy Thursday landed near the present-day city of Veracruz. The Aztec ruler Montezuma thought the Spanish adventurer was the god Quetzalcoatl, the plumed serpent, and sent messengers to meet him. Cortés fired on them. "A thing like a ball of stone comes out of its entrails," was how the Aztecs saw the harquebus. "It comes out shooting sparks and raining fire." By November the Spaniards had arrived at the dazzling island city of Mexico—Tenochtitlán.

"Gazing on such wonderful sights," wrote the soldier Bernal Díaz, "we did not know what to say, or whether what appeared before us was real, for on one side, on the land, there were great cities, and in the lake ever so many more, and the lake itself was crowded with canoes, and in the Causeway were many bridges at intervals, and in front of us stood the great City of Mexico, and we—we did not even number four hundred soldiers!"

By the following June, Montezuma was dead. The Spaniards claimed he died in a hail of stones thrown by his own people, but perhaps they killed him themselves. They were besotted with his treasure. The Florentine Codex, a twelve-volume history from the sixteenth century that contains native Mexican accounts of the conquest, describes Cortés and his Nahua lover, La Malinche, entering the royal treasury.

> [W]hen they entered the house of treasures, it was as if they
> had arrived in Paradise. They searched everywhere and
> coveted everything, for, yes, they were dominated by their
> greed. Then they took out all of the goods which were [Mon-
> tezuma's] own exclusive possessions: his personal belongings,

*all of which were precious: necklaces with thick stones, arm
bands of quetzal feathers, bracelets of gold, golden bands
with shells for the knees, ankle bracelets with little gold
bells, and the royal crowns and all the royal attire, without
number, everything that belonged to him and was reserved
to him only.*

The news of such riches intoxicated Spanish adventurers, including an illiterate desperado named Francisco Pizarro. He was the bastard son of a Spanish colonel and an impoverished rural woman. His mother, probably a maid, abandoned him on the steps of a church. Raised by relatives, he was a distant cousin of Cortés. In 1513 he accompanied the explorer Vasco Núñez de Balboa to the Pacific. Six years later he betrayed his patron, arresting Balboa as he rode to a meeting with his enemy, a rival who accused him of crimes against the king. Balboa was beheaded, protesting his innocence to the end. Pizarro was rewarded with Panama City, which he ran until 1523. The next year he made his first expedition in search of the great civilization rumored to exist to the south. Two years later, he made his second. For demonic speed and horror, few stories can match what happened next, as Pizarro and his tiny band rode into the mountain fastness. It was the lost kingdom of a fairy tale—fantastic, remote, strange, beautiful, and rich. Pizarro was its nemesis, the herald of a greed that his victims could not even comprehend. With Pizarro, a whole continent's appetite for gold rode into the Andes.

THE NAME INCA REFERS TO both the people and the leader of the empire of Peru. Inca expansion from the valley of Cuzco began about 1438, when a talented war leader named Pachacuti Yupanqui conquered his neighbors. He was the first Inca to extend his power beyond Cuzco. Pachacuti was the builder of Machu Picchu, the city that hangs on a mountain ridge forty-three miles northwest of Cuzco, and was his private estate. Pachacuti's son, Tupac Yupanqui, continued the expansion, and by 1471 ruled an empire that extended through the Andes for 3,000 miles.

The culture and appearance of the Andean people evolved through centuries of life in the mountains. They had large lungs and chests for breathing the thin air of the high altitudes. Their steeply terraced fields blazed with lupines, a food crop. Aqueducts carried water to the cultivated land. The Incas had no money. They worked as part of a system of reciprocal duty marked by celebrations that included getting drunk on chicha beer. Wealth lay in the control of labor, the ownership of land, and the state llama herds. The empire was isolated on one side by the planet's greatest ocean and on the other by its biggest forest. In the south there was only desert and to the north they had conquered everyone. The Inca's generals could marshal hundreds of thousands of fighting men. They could move their armies with amazing swiftness along 14,000 miles of roads that astonished the first Europeans to see them. "Such magnificent roads could be seen nowhere in Christendom in country as rough as this," wrote Hernando Pizarro, the conqueror's brother. Another Spaniard described a terrifying climb up a "stupendous mountain-side. Looking at it from below, it seemed impossible for birds to scale it by flying through the air, let alone men on horseback climbing by land. But the road was made less exhausting by climbing in zigzags." The main royal road followed the Andes from Colombia to Chile. A coastal highway paralleled the Pacific, and connecting roads joined the two routes.

Important constructions such as Sacsayhuaman, the immense

fortress above Cuzco, used monolithic stones fitted together with impenetrable skill. They had textiles and ceramics and bronze metallurgy, precious feathers, silver, and rare stones. They mined gold and made gold ornaments. Their surgeons could drill holes in the skull to relieve the fluid caused by head wounds. The survival rate was 80 percent.

They believed they had conquered the whole civilized world. Pizarro went looking for them.

Early in 1527 a Spanish ship was coursing down the west coast of South America when it spotted a sail. It changed course and overhauled the craft, a balsa raft. The boat had cotton sails and an advanced design. Of the captured vessel's crew of twenty, eleven threw themselves into the sea and drowned. The rest submitted. "They were carrying many pieces of silver and gold as personal ornaments," said the Spanish report, "including crowns and diadems, belts and bracelets, armour for the legs and breastplates; tweezers and rattles and strings and clusters of beads and rubies; mirrors decorated with silver . . . emeralds and chalcedonies and other jewels and pieces of crystal."

The ship rejoined its consort, a vessel carrying expedition commander Francisco Pizarro. His men were in a pitiful state, three dying every week from hunger or disease. The coast was barren or impassable mangrove. Pizarro gave his men the option to go home, and most of them took it. Thirteen remained.

The next year Pizarro and this remnant sailed on and reached a coastal town, Tumbrez. From there they made excursions further south until they understood what they had found. They had "glimpsed the edges of a great civilization," wrote John Hemming in *The Conquest of the Incas*, "the product of centuries of development in complete isolation from the rest of mankind."

Pizarro returned to Spain to raise an expedition. He found the court dazzled by the latest treasure from Mexico. The queen gave him a charter of conquest. He returned to Panama, took ship, and on September 24, 1532, after long voyages and persistent sickness, after desperate marches and misadventure, sixty-two horsemen and 106 foot soldiers turned their backs on the Pacific and struck away into the Inca empire.

Tawantinsuyu, as the Incas called their country, stretched 3,000 miles from the center of modern-day Chile to Colombia. The Spaniards marching into it could not have conceived its size. The Inca ruled a population of 10 million. Runners brought news from different parts of the empire, racing in relay along the stone-paved roads. Generals rode in golden chairs carried by liveried servants. Stone silos and warehouses held the empire's food store. The mountainsides shot up like walls.

Tawantinsuyu meant "four parts together"—a federal state bound into a unified whole. But it was a fractious realm, and the Inca spent heavily on gifts to keep the leaders of conquered people loyal to his rule.

The Inca's ancestors also needed gifts. A dead Inca did not lose the prerogatives of state. Each had his own palace in the imperial capital of Cuzco, complete with gold ornaments and decorations, earplugs and feather cloaks and coronets and jade. Treasure chests bulged with their possessions and courtiers waited on them. The ruling Inca paid for it all. He conducted wars of conquest to get the wealth that he had to distribute to maintain his authority.

The throne did not pass from one Inca to the next according to strict rules of inheritance. Wars of succession among competing members of the royal family could divide the country. Pizarro marched into Peru at such a moment.

A powerful Inca, Huayna Capac, had died in about 1527, after a thirty-year reign, leaving two heirs. One son, Huáscar, took the throne in Cuzco while his brother Atahualpa, who had been campaigning with his father, remained in command of the main armies in the north. Tensions drove the brothers into civil war. Huáscar attacked northward, but Atahualpa's general defeated him, capturing Cuzco in 1532, killing Huáscar's family and taking the Inca prisoner. This was the moment that Pizarro began his march into the empire.

It is hard not to think that Atahualpa would have annihilated the Spaniards if he hadn't been preoccupied by war. He knew they were coming. He had sent them gifts. But at the time, he did not even know how the battle for Cuzco had gone. He did not know whether his forces or Huáscar's had prevailed. This was lucky for Pizarro, because Atahualpa may have had it in his mind to enlist the Spaniards against his brother. That would explain why the Inca had not attacked them on their march, when they were most vulnerable. Another break for Pizarro was that the Inca's camp, which might have been months away across the mountains, was close at hand, in the direction of his march, making it likely that he would reach Atahualpa before the news from Cuzco did.

Now the Spanish force entered the shadow of Inca military power. As Hemming imagines it, they climbed an Inca track along a valley that became a canyon. They advanced into the mountains. Everywhere along their route they glimpsed the Inca military outposts high above them, certainly surveying their march.

"The road was so bad [for horses] that they could very easily have taken us there," Pizarro recorded, "or at another pass which we found between here and Cajamarca. For we could not use the horses on the roads, not even with skill, and off the roads we could take neither horses nor foot-soldiers."

On November 15, seven weeks after they had left the coast, the Spaniards reached a pass above Cajamarca, and the sight they saw as the valley opened before them must have frozen their hearts. On a hill above the town spread the Inca's camp. The ruler was in the midst of an army of 80,000. The Inca's tents made a splendid sight, "like a very beautiful city," one of Pizarro's men wrote. "Nothing like this had been seen in the Indies up to then. It filled all us Spaniards with fear and confusion. But it was not appropriate to show any fear, far less to turn back. For had they sensed any weakness in us, the very Indians we were bringing with us would have killed us. So, with a show of good spirits, and after having thoroughly observed the town and tents, we descended into the valley and entered the town of Cajamarca."

So began one of the strangest motions of enemy forces in history. On one side a god-king with 80,000 battle-hardened troops; on the other, a handful of aliens, some of them sick, thousands of miles from any support. Atahualpa could not have thought of such a paltry force as a threat to his person, let alone his empire.

Pizarro sent a deputation to the Inca's camp. They rode through the silent, watching army. Atahualpa was at the hot springs with his women, surrounded by lords and generals. He sat on a low, gold stool. On his forehead was the scarlet tassel that distinguished him from all others. In Spanish accounts Atahualpa displayed unshakable aloofness, neither looking at the Spaniards nor even shifting his position when one rider approached so close that the breath from his horse's nostrils shivered the scarlet threads on the Inca's face. Atahualpa did not look up until he learned through an interpreter that it was the white leader's own brother, Hernando Pizarro, who had come. Making extravagant promises of friendship, the delegation begged the Inca to visit Pizarro in the town. Atahualpa agreed to come the next day. The horsemen returned to Cajamarca to lay their plans.

Pizarro's men spent an uneasy night. The campfires of the Inca's army blazed for miles along the hillside. Fearful of a night attack, Pizarro disposed his men. Infantry and horsemen took up positions in the alleys off the plaza. Pizarro placed four cannon and some musketeers in the little fort along one side of the square.

In the morning a messenger arrived to say that the Inca would come with his men armed. Pizarro replied that Atahualpa would be welcome, as he was Pizarro's "friend and brother." At noon the Inca host moved into the plain, deployed, and waited for the emperor. The Spaniards stayed hidden. The younger of Pizarro's brothers, Pedro, said that he saw "many Spaniards urinate without noticing it out of pure terror."

Atahualpa came in state. His attendants glittered in gold and silver ornaments. A phalanx of court servants in checkered livery went ahead and swept the ground, bending to remove every shred of straw from the Inca's path. Crowds along the way sang as he approached. Half a mile from town the procession halted. The Inca had decided to break his short journey. It was already late afternoon. Soldiers began to set up camp in a meadow. Pizarro, still fearing that the Peruvians meant to make a night attack, sent a messenger pressing the Inca to come to him that day. Atahualpa agreed, and in the last, slanting light of afternoon, the sovereign's escort took the road into the jaws of history.

Most of the army remained behind. About 5,000 lightly armed warriors accompanied Atahualpa's train. The Inca rode on a silver litter carried by eighty nobles in blue robes. His throne was solid gold. Platoons of men wearing gold and silver ornaments marched after him. Atahualpa wore a crown and an emerald collar, his hair intertwined with gold. His litter was thick with parrot feathers and flashed with gold and silver decoration. Sparkling and splendid, in the midst of a brilliant company, the emperor reached the central plaza. He halted in the middle of the square. His standard was planted on a lance. The square, stiff, yellow flag bore the emperor's device: a rainbow, and the Inca's scarlet tassel flanked by upright snakes. Atahualpa looked around for the Spaniards and couldn't see them. "Where are they?" he called out.

At this, the Dominican friar Vicente de Valverde emerged from Pizarro's place of concealment and approached the Inca with a prayer book in his hand. He invited Atahualpa to come inside to meet the governor, as the Spaniards styled Pizarro. The Inca refused, and demanded the return of everything the Spaniards had stolen since entering his realm. Valverde launched into a formal oration known as the Requirement: a proclamation that the Spanish royal government insisted the invaders read out before killing could begin. The priest held out his book and said through his interpreter that it contained the faith he had been sent to bring to the Inca and his people. Atahualpa examined the book, admiring it as an object. Then he hurled it away indignantly, "his face a deep crimson." The friar ran off shouting and weeping. In their memoirs, the Spaniards gave different versions of what the priest had yelled, but whatever his words, they were the signal that Pizarro had been waiting for. The hidden cannon fired into the packed ranks of the Inca's followers and the cavalry burst from concealment.

The Spanish had polished their armor and hung rattles on their chargers for maximum terror. The horses were terrifying anyway, beasts far larger than any in the Andes, snorting and whinnying as their riders rode them into the packed retainers and mowed the lightly armed men down with Spanish steel. The horsemen went through Atahualpa's soldiers like a column of tanks. Pizarro urged his horse into the thick of the fray and reached the Inca's litter, seizing Atahualpa by the arm and trying to drag him off. The Inca's high position, elevated on the shoulders of his retainers, saved him. Pizarro's lieutenants hacked the arms off some the litter-bearers, who still supported Atahualpa's platform on their shoulders. Finally, in the blood-spattered melee, the Spaniards toppled the litter and the Inca fell into the mass. His retainers would not leave him. Every one of them, all men of high rank, died in the slaughter.

As Pizarro was capturing the Inca the massacre proceeded. Packed into a confined space with narrow gates, the native soldiers had no way to escape. Some thousands threw themselves against a wall and broke it down and fled into the plain. The horsemen rode them down, lancing them as they ran and searching for any who wore the Inca's livery. Even after nightfall, according to Spanish accounts, Pizarro's men were spearing Indians. The cavalry thundered on the dark plain, riders and horses like centaurs plastered in blood. To the Peruvians they seemed unkillable. They returned to the square only when Pizarro ordered the trumpeter to sound the recall.

Some Spanish writers claimed that 6,000 Peruvians were butchered in that single bloodbath. "Atahualpa's nephew wrote that the Spaniards killed Indians like a slaughterer felling cattle," Hemming said. "The sheer rate of killing was appalling, even if one allows that many Indians died from trampling or suffocation, or that the estimates of dead were exaggerated. Each Spaniard massacred an average of fourteen or fifteen defenceless natives [a total of 3,000] during those terrible two hours."

As his people were falling in bloody heaps the Inca was marched under guard to the temple on the square. Pizarro ordered clothes to replace Atahualpa's ripped garments, torn in the struggle. Pizarro treated his prisoner with respect. The Inca asked if they were going to kill him, and Pizarro said no—Christians did their killing in the heat of battle, but not in cold blood. They asked their captive how a veteran campaigner could have fallen into so obvious a trap. It emerged that Atahualpa had received bad intelligence about the Spaniards' fighting qualities. He had not imagined that they posed a threat to him, a victorious king with 80,000 men. He had meant to capture them, kill a few, and castrate the rest. All he wanted were the horses.

WITH THE INCA IN SPANISH hands, so was his empire. Pizarro had allowed Atahualpa to send news that he was alive. A divine person even in captivity, the Inca remained in command of his army and people. This authority empowered Atahualpa to conduct the negotiations for his own release, and he set about it. The Spaniards, he perceived, had an appetite for gold. He could not have fully understood it, because the Incas had no money. They valued gold for the way it could be worked. It had its place in the adornment of nobles and in sacred rites, but even in those functions gold was not the top material. Jade and some kinds of feathers outranked it. Yet gold was what the white soldiers wanted. The Inca decided he would offer them all they could possibly want, believing that would make them go away.

In Cajamarca today you can see the room that Atahualpa offered to fill with gold in exchange for his freedom. It measures seventeen feet by twenty-two feet. The Inca offered to pack it with gold in two months to the height of his raised arm. Pizarro accepted. A notary recorded the details of the bargain. The Inca's couriers fanned out into his domain and a stream of objects flowed back along the royal roads to Cajamarca. There were cups and vases and delicate statuettes of llamas, gold birds, and trees and a device like a fountain spewing gold. The amount of gold available staggered the Spaniards. A solid-gold sacrificial altar in Cuzco weighed at least half a ton. A golden fountain weighed more than 700 pounds. In a sanctuary, Spanish soldiers found an old woman in a gold mask fanning flies from the remains of dead Incas. The rulers' treasure lay there for the taking.

The wonders dazzled the Spaniards. They described little gardens made of gold, in which every clod of earth and cob of corn was solid gold. There were gold chalices set with emeralds. Andean goldsmiths had been making objects for a thousand years, since about 500 BC. Some of the articles collected dated from even earlier, and showed Chinese and Vietnamese influence, suggesting that Asian voyagers might have reached Peru a millennium before the Spaniards. Pizarro kept a few of these objects to dazzle the Spanish court. They melted all the rest. In a single month they retrieved the equivalent of 1,326,539 gold pesos, or about $340 million in today's money. The booty weighed five tons, not including the 190-pound solid-gold throne. Just that one month's bullion harvest equaled a full year's output of European gold. It would never buy the Inca's liberty.

In Atahualpa, Pizarro had the tap to drain Peru of gold. The Inca's person was sacred. Even in his prison room, he was surrounded and served only by his wives and sisters. They fed him from their hands. The Inca wore clothing softer than anything his captors had ever seen. He had a cloak made from the skins of vampire bats. When the Inca spat, a woman caught the saliva in her hands. Everything he touched—rushes he walked on and the skeletons of birds he ate—was packed into leather chests and later burnt, the ashes flung to the wind to prevent anyone from touching what the Inca had touched. When great lords visited him, they kissed his feet and hands. The Inca did not look at them. Because of the Inca's divinity, his orders to bring gold to Cajamarca were obeyed; it did not matter that he was a prisoner.

The captive Inca ruled through his generals. Of the army at Cajamarca, about 75,000 were still encamped above the town. Another 35,000 troops under Atahualpa's commander in chief stood between Cajamarca and Cuzco. The general who had captured Cuzco from the Inca's brother had a further 30,000 troops to garrison the capital. Against such numbers the asymmetric advantages of hard steel and horses would not have saved the Spaniards had the Peruvians attacked en masse. They did not attack because the Inca told them not to. That is why a small party of horsemen under Pizarro's brother could ride from Cajamarca to Cuzco, strip 700 four-and-a-half-pound gold plates from the sun temple, and return to Cajamarca without harm.

On April 14, 1533, long-awaited Spanish reinforcements reached Cajamarca from the coast, to the jubilation of the garrison. Pizarro had been waiting for the 150 soldiers. Their arrival almost doubled the occupying force. It seems absurd to speak of 300 men "occupying" a mountain empire 3,000 miles long, where vigorous officers commanded more than 100,000 troops. Yet the Spaniards controlled the god-king, and he controlled the men.

Atahualpa grew fearful when he learned of the arrival of reinforcements, for he took it to mean that the Spanish planned to stay. The ransom of gold would not suffice to make them go away, as he had believed. He clutched at the straw that they would honor their bargain and set him free.

The treasure was pouring into Cajamarca. Pizarro had nine forges melting about 600 pounds of gold a day into ingots. They stamped the newly smelted gold with the Spanish royal mark and added it to the growing hoard. If gold had retained any of its sacred status, the Spaniards extinguished it at Cajamarca. The artistic output of a thousand years vanished into the furnaces. It must be one of the most potent images in history—the transformation of a culture into cash. On one day alone, a train of 235 llamas wound into Cajamarca laden with gold and silver. Each Spaniard was entitled to a share agreed upon and stipulated, and clerks recorded every ounce. In the end, a foot soldier would get about forty-five pounds of gold and twice that weight in silver; a horseman, double those amounts. Francisco Pizarro had about 630 pounds of gold and half a ton of silver coming to him, plus Atahualpa's throne as a bonus. The king was entitled to a fifth of everything—the royal *quinto*. Spanish court officials were on hand to see that he got it. Everything was entered in a ledger. To the scratch of pens the Inca's patrimony went into the furnace and a river of gold flowed off to Europe. It bought the Inca nothing.

On June 12, Hernando Pizarro, the older of the governor's two brothers, left Cajamarca, and Atahualpa wept. The Inca perceived the departure as the loss of a protector. Hernando had assured the ruler that he would not let him be killed, and Atahualpa had believed in his sincerity. Now that protector was gone, and the Inca's fate approached.

Pizarro heard that the Inca general Ruminavi had raised an army of 200,000 and was hurrying south from Quito to attack the Spaniards and restore the Inca. It is possible that Atahualpa had, as Pizarro suspected, ordered his general to rescue him. John Hemming thought it could not be proved one way or the other. In Cajamarca the mood of the Spaniards was near panic. A strong faction clamored for the execution of the Inca for "treason." Pizarro is said to have wept at Atahualpa's pleas to be spared. In the end there was no trial, only a meeting of Pizarro's council. They decided on death by fire, and as night gathered on July 26, soldiers led the Inca into the square and tied him to a stake while trumpets blew. A priest explained to

Atahualpa that he could escape burning by converting to Christianity. To the Inca, death by fire was shameful, and he agreed to become a Christian. The priest baptized him Francisco, after Pizarro. Then they strangled him.

Pizarro was vilified for the murder, then redeemed, then denounced again. His band broke up in quarrels about sharing the loot. Few got back to Spain. Some died in fights among themselves; some gambled away their money. Hernando Pizarro returned to Spain in 1540, where his enemies had him thrown into prison. He stayed there twenty years and came out a broken man. Francisco, the great conquistador, was stabbed to death in 1541 by malcontents in Lima, as he sat at his table eating supper. The gold transformed the finances of the West.

3

THE MASTER OF MEN

*In truth, the gold standard is already a
barbarous relic.* —John Maynard Keynes

SPANISH GALLEONS PLOWED ACROSS THE SEA FOR
home in tremendous treasure fleets. Sixty ships strung out for miles
on the blue ocean tacked and plunged as they picked their way
through the reefs of the Florida Strait, the channel that connects
the Gulf of Mexico to the Atlantic. The warm waters could brew
hurricanes. "The sun disappeared and the wind increased in velocity
coming from the east and east northeast. The seas became very giant
in size," a mariner wrote of a storm that wrecked a fleet. They had
more than weather to worry about.

French and English privateers—privately owned warships sanc-
tioned by their governments—attacked the treasure ships. In 1523
the French corsair Jean Fleury seized three caravels loaded with Aztec
gold and jewels and exotic animals, off the coast of Portugal. The

English sea dog Francis Drake spread dismay along the Spanish sea routes from the Caribbean to the shores of Spain. Such raiders helped to execute their countries' foreign policies: containing Spanish power by reducing the wealth that helped Spain pay for European war. The French considered it so vital to raid the Spanish fleets that not even peace with Spain in 1559 could stop them. The parties drew a line through the Canary Islands and agreed that no action west of the line would break the European armistice. There would be "no peace beyond the line."

Even so the great fleets brought a stunning treasure over the sea and up the river to Seville. In 1564 alone, 154 ships reached the port. The tide of bullion almost beggars belief. In a hundred years Europe's gold supply increased by five times. Money washed into the continent. Spain struggled to master this bonanza, and so did Europe. International trade expanded, and the story of gold in human affairs became the story of how to manage the commerce between countries. Whether the famous solution to this challenge—a system called the gold standard—bent gold to our will or bent us to gold's, is the subject of this chapter: a subject that even now, with the gold standard dead in a ditch for more than forty years, still excites rancor.

WHEN THE SPOILS FROM MEXICO and Peru came spilling into the country, Spain was not well equipped to handle the jackpot. There was no native professional class to manage money. Spain had driven out its mercantile class in 1492, with a royal edict expelling Jews and Muslims. The merchants and bankers who replaced them were foreigners, Italians and Dutch, who favored business connections with their own countries. Much of the fresh wealth was funneled through

Spain to other destinations. "Gold and silver merely acquired their international status in Spain," one study said, "without being in any way connected with the Spanish economy."

With the inflow of gold, the continental economy expanded. As commerce grew, so did the international trade fairs where much business was transacted. There was now more money. Scores of gold coins circulated in Europe, all of different value. To handle the exchange, bankers and moneychangers came to the fairs. A merchant could exchange his own money for the money that his counterparty wanted. It was a cumbersome and dangerous system, with sacks of coins carted here and there through Europe. The solution to this unwieldiness was paper money.

Merchants increased the use of bills of exchange, a system that prefigured modern checking. Like checks, the bills were contracts to pay, and bankers would redeem them in cash for a fee. Traders attending fairs did not have to bring money, but could issue paper promises. In the late 1600s this system expanded and got easier to use when London's goldsmith bankers (goldsmiths who had branched into foreign exchange) began to accept each other's paper bills as part of a competition for customers. The use of paper spread from private commerce to public finance when countries issued paper money backed by gold or silver. The circulation of gold coins decreased until the actual exchange of metal marked only the largest transactions, such as clearing trade imbalances between countries. Finally gold's place in international affairs settled into the system called the gold standard.

The gold standard operated by binding countries to a strict agreement: the national currency supply had to match the amount of gold bullion in reserve at a stipulated ratio. If you were a foreigner accepting the banknotes of a gold-standard country, you did so knowing that you could redeem the notes for gold. The store of gold determined exactly how much money a country could have in circulation. It could not print more notes unless it had more gold. Generally, gold-standard fans approve of this, and gold-standard opponents don't. The strict operation of the gold standard sent regular waves of misery through the world, as the vagaries of trade would drain a gold supply and lacerate an economy. The great historian of the gold standard, Barry Eichengreen, believes that the system caused the Great Depression, by preventing the government from stimulating the economy with cash. Nevertheless, a pro-gold sentiment has returned in some quarters.

The president of the World Bank, Robert Zoellick, sent the bullion price spiking in November 2010 with remarks suggesting gold might resume some monetary role. Some central banks have started stocking up on gold after years of selling. On the American right, advocates hark back to an imagined simpler and more upright time. But if the gold standard was simple, it was the simplicity of a war club. Some of the most ruthless passages in history have been set off by a good swing of the gold standard. What's more, even those who agreed that paper money needed the ballast of a metal to control it did not always agree on what that metal should be. Silver was a much more common money than gold, and many countries backed their paper with both.

WHEN SPANISH GOLD WAS SWELLING the European money supply, a silver penny had been circulating in Britain for eight hundred years. The British pound originated as one pound's weight of silver pennies, or 240 of them. Similar silver coins—or even the British coins—circulated in Europe. There was a demand for silver money, and Spain, with an abundant supply, fed the demand from deposits in Bolivia and Mexico. In the sixteenth century such silver coins as the so-called Spanish dollar oiled the wheels of international trade. (With a face value of eight reales, these were the famous "pieces of eight.") In 1785 the United States adopted a dollar as its currency and based it on the Spanish coin. Congress set a silver standard in a 1792 law that spelled out how much silver the mint had to put in every dollar, half dollar, quarter, and "disme"—the old spelling of dime. But the act defined the values of gold coins too, such as the ten-dollar eagle, and specified the amount of gold each coin must contain. Defining the value of currency in terms of two metals put the United States on what is called a bimetallic system.

The challenge of bimetallism is obvious—how to fix a ratio between the two metals. Say the silver price suddenly went viral and eclipsed gold: the metal value of a silver coin could in theory overtake that of a gold coin of higher denomination. The Treasury addressed this problem by fixing the values of the metals. In the United States, the mint price of gold—what the mint would pay for it—was fixed at \$19.3939 an ounce, and for silver \$1.2929, a ratio of 15:1. The problem with fixing a ratio is that events can unfix it.

A declining world gold supply and a European war increased demand for gold. In 1797 the report of a French fleet landing an invasion army in Wales caused a run on the Bank of England. The report was false, but before it could be countered Londoners rushed to cash their gold-backed banknotes into gold. Some £100,000 worth of gold was leaving the central bank every day. To keep the national reserve from evaporating, Parliament passed a law that made banknotes "deemed payments in cash."

The French too were short of gold. In 1803 they sold Louisiana to the United States for $15 million—at 4 cents an acre, a good deal: it doubled the area of the United States. The Americans paid for the land with U.S. government bonds. The bankers running the transaction sold the bonds for cash, and Napoleon got the gold he needed to pursue his war.

Inevitably the appetite for gold pushed up the price, driving it above the exchange rate set by law in the United States. An owner of gold could sell it for more in England than he could in America. In response, gold flowed out of the United States to Europe, toward the higher price. Traders used the higher gold price obtained in Europe to buy silver, took the silver back to the United States, and used it to buy more cheap gold at the established rate. By 1834 the supply of gold in America was so low that a visiting French economist wrote, "Since I have been in the United States, I have not seen there one piece of gold money, except on the scales at the Mint. Once minted, gold is embarked for Europe and remelted." Congress acted, setting the gold price even higher than it was in Europe. The gold-to-silver price ratio became 16:1. Now a trader could get more by buying gold in Europe and selling it in America. The gold flowed back.

The ebb and flow of gold not only described the fortunes of countries; it intensified them. A nation with a trade deficit, for example, would see its gold stock dwindle as it paid out bullion to foreign creditors cashing in their paper money. The shrinking gold stock of the deficit country would curtail domestic spending by the government. Business activity would suffer from the declining money supply.

It was not a scarcity of gold in the world at large that caused these problems. There was soon to be more gold than anyone could have conceived. It erupted into the world from a series of breathtaking discoveries—a gusher of new gold that poured into the markets and founded the modern gold supply. Far from taming the demons of international finance, the gold invigorated them.

ON JANUARY 24, 1848, JAMES MARSHALL panned some bright flakes from the water at Sutter's Mill, on the south fork of the American River, 130 miles northeast of San Francisco. "This day," one of his workmen noted in a diary, "some kind of mettle was found in the tail race that looks like gold." Within days Marshall had his crew knee-deep in the river dredging soils. They tried to keep the find secret, but word got out. In the next seven years, 500,000 men swarmed into the Sierra from around the world.

It was the first gold rush of the modern age. "The whole country from San Francisco to Los Angeles and from the seashore to the base of the Sierra Nevadas resounds with the sordid cry of gold, GOLD, GOLD!" the San Francisco *Californian* reported. "The field is left half planted, the house half built, and everything neglected by the manufacture of shovels and pickaxes." They dammed the streams and sluiced the gravels, they pulled the hills apart. "Three men using nothing but spoons dug $36,000 dollars in gold from cracks in a rock," said one account. "A rabbit hunter poked a stick in the ground, hit rock quartz, and dug up $9,700 worth of gold in three days." But these were the lucky few; the California gold rush didn't run on spoons. A different utensil appeared to exploit the goldfield: the mining company.

Dozens of new companies formed, attracting capital from the cities of the east coast and Europe. When silver was discovered in Nevada, the speculative boom caught fire. The dozens of companies became thousands. The inrush of investment supported the development of new technology, but not until later in the rush, when the most accessible deposits had been emptied out. The first equipment in the hills was rudimentary. The sluices were no different in design from the ones the Romans had used. Water washed away light soils to reveal the heavier flakes of metal left behind. The chain pumps brought to California by Chinese miners had been known in Asia from antiquity: a treadle turned a belt that carried wooden trays: the trays scooped water out of flooded areas and fed the sluices.

Tradition bathes the California gold rush in a honeyed light. "Few events in the history of the United States have been as glamorized," one expert on the period said. "From the nineteenth century to the present, most historians have portrayed it as both a heroic and dramatic epic and as a giant step toward the fulfillment of the nation's Manifest Destiny." In this rousing view, a horde of enterprising men catch fortune's wind and write the founding story of the Golden State. The truth is bleaker. Some Americans brought slaves to the mines. One researcher reckons that half the black men working in California in 1850 were slaves. Mexican miners lived in peonage, tied by debt to their *patrones*. Chinese "coolies" worked as indentured labor under a "credit-ticket system" that bound them to the middlemen who'd paid their passage. White Americans weren't slaves, but they toiled from 6:00 A.M. in freezing water that flowed down from the melting snow at higher elevations. A miner had to wash 160 pails of dirt a day to get an ounce of gold. "You can scarcely form any conception of what a dirty business this gold digging is," one miner wrote. "We all live more like brutes than humans." Even the investors suffered, as insiders manipulated the stock market, cheated shareholders, wrested mines from weaker individuals, and drained the cash from companies into their own pockets.

But the gold billowed into the world. The satanic mill of slavery, oppression, and fraud was a bullion spinner. In terms of today's money, billions of dollars worth of gold flowed from the hills of California. It swelled supply with almost unimaginable speed. Robert Whaples, a professor of economics at Wake Forest University, has calculated that from 1848 to 1857, California produced 848 tons of gold. At the official U.S. government price of $20.67 an ounce, that single decade's production amounted to a staggering $561 million worth of gold—nearly two percent of the gross domestic product of the whole country. New discoveries elsewhere in the world added even more production, and by 1852, only four years after the discovery at Sutter's Mill, the world's gold mines were producing 280 tons a year—forty times the volume at the end of Spain's century of plunder and 200 times the volume from before it. A cataract of gold went foaming into an eager market.

"As the creditor of the whole earth," wrote one historian, "London got the first of this gold." In four years the Bank of England grew its gold reserves from £12.8 million to £20 million. The Bank of France bought even more, increasing its bullion stock from £3.5 million to £23.5 million. At the time, only Britain was formally on the gold standard, with banknotes convertible to gold. But the flow of so much gold into the financial system, and Britain's place as world banker, opened a crack in the continent's facade of bimetallism. In 1871 Germany bought £50 million worth of gold and issued a new gold-backed currency. "We chose gold," said Ludwig Bamberger, a German politician, "not because gold is gold, but because Britain is Britain."

The silver dominoes began to fall. In the United States, the Coinage Act of 1873 effectively demonetized silver—the "crime of 1873," as the silver lobby later called it. Scandinavian countries scrapped the monetary role of silver in 1874, and Holland followed one year later. France and Spain went on the gold standard in 1876. Looking back, it can appear as if an irresistible monetary wisdom was sweeping the large trading nations into its inevitable embrace. But some scholars think bimetallism was a better system than the gold standard, partly because the use of both metals provided the stability of a bigger money supply. The gold standard did produce periods of financial harmony, but could also suddenly reverse the fortunes of a country.

In 1861, when the costs of the Civil War were swallowing the federal gold stock, the United States suspended dollar redemptions. Instead of a convertible dollar, the Union issued a paper note, the infamous "greenback." But in 1875, as the gold standard was consolidating its rule, the United States reinstituted gold convertibility, and set 1879 as the date for it to begin. As that date approached, however, a trade deficit with Europe meant that America faced the prospect of a rapid outflow of its gold. The Treasury was saved at the last minute by a European disaster: a late-spring frost wiped out French and English crops. Now the shoe was on the other foot.

As the cost of wheat shot up in Europe, America had a bumper crop. Gold moved from Europe to the United States to buy the needed food. So began a three-year boom in American farm exports, and a hefty boost to the gold reserve. But the gold standard operated with perfect impartiality in dealing out misery; the next twitch of international finance would almost ruin America.

In 1890, a banking crisis in Argentina rattled European confidence in overseas ventures. Investors began to sell American debt, exchanging their dollars for gold. Now gold flowed the other way, out of the government's vaults and onto ships and across the ocean to Paris and London and Berlin. By 1892 the American gold reserve had fallen to $114 million, dangerously close to the $100 million level the Treasury had set as its minimum reserve. Once again the United States took the most drastic measure available: it halted gold payments.

The havoc that followed showed how radical and useless such an action was. The public concluded that the whole banking system was circling the drain, and started a run on the banks. The trade deficit with Europe widened to $447 million—a horrifying abyss. The interest rate for short-term money in New York climbed to 74 percent. The stock market fell apart. Four railroads went under, and 500 banks. Thousands of businesses failed. In this tempest of disaster, the United States was facing ruin.

The government had $40 million left in monetary gold, a store that was dwindling by $2 million a day. In the financial community, they thought it certain that the Treasury would fail. On the last day of January 1895, $9 million in gold bullion—almost half the reserve—sat on ships in New York Harbor bound for Europe. At the last possible moment, at the very brink of the abyss, the country was rescued from insolvency by one of the men that the public was blaming for the state of things, the greatest grandee of them all, the American prince—the financier J. Pierpont Morgan.

In American terms, the Morgans had been rich forever. They had not scrabbled their way out of poverty but had glided along a path to prosperity that began sixteen years after the arrival of the *Mayflower* at Plymouth, when Miles Morgan bought a farm at Springfield, Massachusetts, and set about, in the words of the family's chronicler, "spawning generations of land-owning Morgans."

Morgan's father, Junius, moved to Boston in 1851 to expand his business into merchant banking. He enrolled his son at the English High School, admonishing him to make friends with those of the "right stamp." Morgan was a lively, passionate, and moody youngster. He was beset by sudden rashes on his face and suffered bouts of scarlet fever. In 1852 he was sent to the Azores to recover from a rheumatic disease that left one leg shorter than the other.

Banking then was a dynastic occupation. The great English banking families such as Baring and Rothschild fed a cult of personality that Walter Bagehot, a British journalist and commentator of the day, captured when he wrote: "The banker's calling is hereditary; the credit of the bank descends from father to son; this inherited wealth brings inherited refinement."

Junius moved his headquarters to London in 1854. Morgan went to a school on Lake Geneva, and then to the University of Göttingen. Later he returned to America to become his father's eyes and ears on Wall Street.

Troubled throughout his life by his skin, and afflicted at times by nervous ailments and migraine headaches, Morgan nevertheless became the greatest banker in America, a tall, burly, laconic, and intrepid financier, prowling Wall Street behind the smokestack of a huge cigar. His yacht, the *Corsair II,* with its sleek black hull and yellow funnel, was said to be the largest private craft afloat. He collected bronzes, porcelains, watches, ivories, and paintings, rare books, manuscripts, and ancient artifacts. He bought rare furniture, tapestries, and armor. In two decades he spent almost a billion dollars in today's money indulging his passion for collecting. In his refinement, cosmopolitan upbringing, and fortune he was probably the only man in America that European bankers would accept as an equal, and for that reason, he was political poison in the United States.

Still a largely agricultural country, the United States was not the creditor it is today. It was a debtor nation. Its rural voters hated the eastern establishment bankers, whom they viewed as having enslaved America to British gold. The operation of the gold standard could punish farmers by depressing prices, an effect they attributed to the wicked machinations of Europeans, abetted by American financiers. Such was the political temperature at the end of January 1895, as $9 million sat in New York waiting to be shipped to Europe.

As the sense of crisis heightened, Morgan held a meeting with August Belmont, Jr., the Rothschild agent, at the New York Subtreasury. Just the fact of the meeting of two such powerful financiers drained some of the tension from the situation. Overnight, the $9 million in bullion was taken from the ships and put back in the government's vault. But the relief was temporary. In Washington, the cabinet rejected a private bond offer put together by the two banking houses, and gold started leaving the Subtreasury again as skittish dollar owners cashed in their paper. Morgan boarded his private railway car and set out for Washington.

On arrival he went to the White House. He was told that the president, Grover Cleveland, would not see him. Morgan replied: "I have come down to see the president, and I am going to stay here until I see him." He stayed in the White House all day. He returned to the Arlington Hotel, played solitaire all night, and in the morning walked back across Lafayette Square, and was shown into a meeting in the president's office. He sat there silently while Cleveland and members of his cabinet discussed the emergency. Finally a clerk came in and informed the secretary of the Treasury, John G. Carlisle, that the government was down to its last $9 million in gold coin. Morgan spoke up, informing Cleveland that he knew about a $10 million draft soon to be presented. "If that $10 million draft is presented, you can't meet it," he said bluntly. "It will all be over before three o'clock." Cleveland made the only sensible reply he could: he asked Morgan what to do.

Morgan's solution was his masterpiece—a $65 million bond to

be sold to a European syndicate organized by Morgan and the Rothschilds. Once the important European banking houses joined the syndicate, New York banks came in too. In exchange for a premium interest rate, subscribers agreed to pay for the bond in gold. Not only that, they promised to "exert all financial influence . . . to protect the Treasury of the United States against the withdrawal of gold" until the bond was paid off. It was a brilliant stroke. Morgan had kept America on the gold standard by suspending it. Without the need to redeem foreign-owned dollars in gold, the American treasury had time to restore itself.

FOR ALL ITS HARSHNESS, THE gold standard reigned over a period of economic expansion. Its gift to the industrializing and trading world was the credibility of one another's currencies. The system solved two main problems.

First, it removed uncertainty about fluctuations in the value of a currency. With a currency defined in terms of gold, the holders of the currency could make rational decisions about the future, because they knew what the currency would be worth at any time. If they owned dollar bonds maturing in twenty years, say, they knew what the bonds would be worth because that value was expressible in terms of gold.

Second, the gold standard told central bankers what to do with monetary levers such as interest rates. Let's say the central bank's interest rate was low. Money poured out into the economy as borrowers took advantage of easy credit. The fresh money stimulated the economy, which was the object of the low interest. But when was the stimulus just right, and when too much?

On the gold standard, a central banker could see when prices in the stimulated economy rose too high, because the rise in prices made gold look cheap. Noticing this, dollar owners would start converting their currency into gold, taking advantage of the bargain. Gold left the Treasury, and in came paper money. But according to the gold standard, the amount of gold and the number of dollars had to tally at a certain ratio or there would not be enough gold to back the dollar. To stop the flood of paper money in and gold out, the central banker would have raised interest rates. Suddenly you could earn more by cashing in your gold for dollars, and investing the dollars in the economy. Gold flowed back in. Once things had matched up in the Treasury again, the central banker could ease off on interest, and so it went.

Proponents of a return to the gold standard are seduced by the apparent serenity of this monetary picture, and either forget, or think we should accept, the bloodshed in the background. Those who disagree with them consider gold a blunt instrument for monetary purposes today, when economic planners have an abundance of data, such as the consumer price index, to help them assess how a currency is doing. Against these more sophisticated measurements, then, gold is just not a good indicator.

There are other reasons too that economists think gold's hour has passed. In a growing economy, for example, a gold-backed currency has to cover a correspondingly growing number of transactions, and it can only do this if prices fall, which is deflation, a killer of jobs.

"Under a true gold standard, moreover," writes Barry Eichengreen, "the [Federal Reserve] would have little ability to act as a lender of last resort to the banking and financial system. The kind of liquidity injections it made to prevent the financial system from collapsing in the autumn of 2008 would become impossible because it could provide additional credit only if it somehow came into possession of additional gold. Given the fragility of banks and financial markets, this would seem a recipe for disaster. Its proponents paint the gold standard as a guarantee of financial stability; in practice, it would be precisely the opposite."

THE GOLD STANDARD WAS THROTTLED to death on live TV on a Sunday night in Washington. Its crime: hamstringing the government into whose care it had been placed, that of the United States. In some ways the situation that led to the system's demise was a replay of the 1890s—a lousy American balance of payments and the sucking sound of large amounts of bullion leaving the Treasury. Yet in the intervening years the condition of the United States had changed almost beyond belief. It was the world's titan. Five years after the end of World War II, two thirds of all the monetary gold reserves in the world belonged to the American government. It had 20,000 tons of bullion in its deep-storage vaults, such as the one at Fort Knox, Kentucky. Why would Americans, with such a gold position, sideline the power that the metal represented?

4

CAMP DAVID COUP

I have directed the Secretary of the Treasury to take the action necessary to defend the dollar against the speculators.
—Richard Nixon, address to the nation, August 15, 1971

As with any satisfying murder story, we must set the stage for gold's last moment at the center of monetary life. We begin the story where so many really first-rate crimes have taken place—London. It is the Great Depression. Gold is surging out of the Bank of England as foreigners, uneasy at Britain's prospects, take the cash and run. Things are so bad that King George V takes a £50,000 pay cut. When the government reduces naval pay, sailors at a base in Scotland go on strike. To foreign depositors, the word "mutiny" signals that a revolution is at hand. Forty million pounds in gold flies out of the central bank in a week. The Bank of England begs the government to allow it to stop redeeming notes in gold. Parliament passes the necessary legislation. Britain is off the gold standard!

Within days, forty-seven countries followed suit. A year later, only five countries were still on the gold standard: France, Switzerland, the Netherlands, Belgium, and the United States. In the mood of panic, redemptions into gold began to churn through the system. In a single day the Belgian central bank took $106 million in gold from the U.S. Treasury and France took $50 million. A few weeks later the French came back for $70 million more. President Herbert Hoover warned that all the bullion "dashing hither and yon from one nation to another, seeking maximum safety, has acted like a cannon loose on the deck of the world in a storm."

The man who would stop the rout of America's reserves trounced Hoover in the election of November 1932. In the practice of the time, Franklin D. Roosevelt, the victor, would not be inaugurated until the following March. In the lame-duck interregnum, much could happen . . . and did. The cannon that Hoover had identified kept smashing around on the deck. When one of the men tipped for Roosevelt's cabinet let drop that it might be smart to quit the gold standard, a run on the dollar took $320 million out of the Treasury in less than two months. Worse, some of those fleeing were Americans. Distrustful of Roosevelt, who refused to say what he planned to do, they cashed in their currency for gold. It wouldn't do them any good.

Roosevelt took office March 4, 1933. On March 9 he rushed through the Emergency Banking Act. The act gave the president the power to regulate ownership of bullion. On April 5 he signed Executive Order 6102 "forbidding the Hoarding of Gold Coin, Gold Bullion, and Gold Certificates within the continental United States." The order made it a crime for any individual, partnership, association or corporation to possess gold. The law exempted "customary uses" such as gold fillings for teeth, and rare coins in collections. All other gold was the government's, and owners had to deliver it by May 1, twenty-five days after the signing of the order, when they would be paid the official government price of $20.67 an ounce.

A challenger immediately tested the law. Frederick Barber Campbell, a sixty-seven-year-old New York insurance lawyer, marched into the Chase National Bank and asked for the twenty-seven gold bars that he had there on deposit. The bullion was worth $200,000, a large sum. In accordance with the presidential order, the bank refused to surrender the gold, and Campbell sued them. He also asked the court to restrain the bank from delivering the gold to the federal government. Campbell maintained that the order, and the law cited by the president as his authority for it, were both unconstitutional. The government responded by indicting Campbell for failing to report his holdings to the Treasury, and then indicting him again for hoarding. If convicted on both counts, Campbell faced a maximum term of twenty years. In the end he did not go to prison, but his lawsuit failed. Campbell died three years later of a heart attack, in the Metropolitan Club on East 60th Street, where he lived. The Treasury got the gold.

As private gold shifted to the government, the public coffers swelled. In 1930 the United States had about 7,000 tons of gold. In 1935, two years after the confiscations started, the stock reached almost 10,000 tons. Not all of this increase came from private gold. As Europeans grew fearful of approaching war, some shifted wealth across the ocean to the safe haven of the United States. This increased the bullion flow into the Treasury. When war broke out in 1939 the United States, with its industries intact and still two years away from war, became an important supplier of arms and other goods to combatants. Gold imports soared. In 1940 the Treasury had 21,000 tons, a five-year increase of more than a hundred percent. Nor was the size of the reserve all that had changed. The cost of an ounce of gold had climbed from $20.67 to $35, a hike of almost 70 percent. It had happened very fast, and for one reason—the United States had been cooking the price.

As soon as he had issued the gold confiscation order, Roosevelt obtained the authority to reduce the amount of gold a dollar represented by as much as 60 percent. It would take more dollars to buy an ounce of gold. The dollar would be worth less than it had been, but there would be more of them, and that's what Roosevelt wanted—more currency in circulation to stimulate the economy out of the Depression.

Raising the gold price was easy: the Treasury just bid it up. Henry Morgenthau, Jr., Roosevelt's treasury secretary, recorded in his diaries how he and the president rigged the price.

> Franklin Roosevelt would lie comfortably on his old-fashioned three-quarter mahogany bed. . . . The actual price [of gold] on any given day made little difference. Our object was simply to keep the trend gradually upward. . . . One day, when I must have come in more than usually worried about the state of the world, we were planning an increase of from 19 to 22 cents. Roosevelt took one look at me and suggested a rise of 21. "It's a lucky number," the president said with a laugh, "because it's three times seven."

Preserved throughout the war, the huge gold reserve of the United States gave it a uniquely powerful hand to shape the world that followed, and in a feat of statecraft pulled off in the mountains of New England, to obtain for the dollar a status that has guaranteed the U.S. Treasury a cheap supply of money ever since.

· ● ·

IN JULY OF 1944, ONLY weeks after D-Day, as Allied armies battled out of Normandy in the invasion that would defeat Germany, the representatives of forty-four countries arrived at a meeting thousands of miles away to plan for the victory. At the town of Bretton Woods in the New Hampshire forest, sleek black cars swept up the drive to the regal Mount Washington Hotel. Within days, 730 delegates were rewriting the rules of international finance. An American participant sounded a rousing theme. "We fight together on sodden battlefields," said Fred M. Vinson, later chief justice of the United States. "We sail together on the majestic blue. We fly together in the ethereal sky. The test of this conference is whether we can walk together, solve our economic problems, down the road to peace as we today march to victory."

They would go down the road to peace, all right, guided along it every step of the way by the United States. America and Britain seemed to share the decisive power at Bretton Woods—the United States because it was rich and militarily preeminent; Britain, because its empire was so vast. But the British power was illusory. A newly discovered transcript of the Bretton Woods meetings shows the empire "disintegrating before your eyes," one expert who read it said. Britain, shattered by war and in debt, followed the course the Americans wanted, and although they often objected, so did everyone else. "Now the advantage is ours here," Morgenthau said to a member of his team, "and I personally think we should take it."

They did take it. There was no question of a return to the old gold standard system, because the United States had 75 percent of the world's monetary gold. At Bretton Woods the participants agreed to peg their currencies instead to the U.S. dollar, and the dollar would be convertible to gold. The system enshrined the dollar as the world's reserve currency, giving it advantages it still enjoys. Most commodity prices are denominated in dollars, and Americans buying such commodities escape the transaction cost that other countries' citizens must pay to get the dollars they need to make their purchases. The international community's need for dollars also means that the Treasury has a ready market for its debt, and generally can pay a lower interest on it.

The United States was a cornucopia of what the world needed as it recovered from war, especially capital. America wanted open markets and free trade. Two of the institutions created at Bretton Woods, the International Monetary Fund and the World Bank, helped the United States achieve those ends. Dollars flowed into the war-torn countries, and their economies began to recover. As Americans imagined it, the flow of money out of the United States would in time be matched by money flowing back in, as the reconstructed national economies gained the ability to buy American products.

For a while, that's how it did work, with Japan and Europe "eager for dollars they could spend on American cars, steel and machinery." But beginning in 1950, the U.S. share of world output dropped from 35 percent to 27 percent. Also, American spending, especially on the Vietnam War, deluged the world with dollars. In one indicator of this trend, foreign-owned dollar deposits in U.S. banks and foreign-owned purchases of U.S. government debt more than doubled between 1950 and 1960, from $8 billion to $20 billion, and that was before the Vietnam spending took off. Other countries' citizens were not spending the money as fast as they were banking it. Americans bought the goods produced in the new factories of other countries, factories often built with American capital, but those countries did not buy as much from the United States. Surplus dollar holdings built up in the central banks of foreign countries. In the gold-standard regime, that would signal the balancing of books. Bullion would leave American ports for European and Asian destinations. And that's what happened.

At the center of the Bretton Woods system lay that old serpent, gold. While the main financial action ran on dollars, gold convertibility still backed those dollars. In the first postwar years, because the currency was what everybody wanted, dollar convertibility was essentially a technicality. This technicality changed to practice with crushing speed. In a span of only thirteen years, the combined effect of international largesse and foreign war, a global military establishment and trade deficits, drained the Treasury of the United States of half its gold. Moreover, the number of foreigners holding U.S. government debt had doubled. If foreign creditors had all at once cashed in their dollars for gold, the U.S. gold stock would have vanished. Only the fear of provoking a run on the dollar and undermining their own dollar holdings prevented foreign owners of U.S. banknotes from cashing them in. But the world was getting edgy. At a point when the United States owed $60 billion and its gold stock was $12 billion, a French economist pointed out that asking America to pay its debts in gold was pointless: "It's like telling a bald man to comb his hair," he said. "It isn't there."

In early 1965 the French president, Charles de Gaulle, fuming at America, told a press conference at the Elysée Palace that the dollar had lost the right to be the world's money, and the international community should return to the only standard that made sense. "Gold!" he intoned, the standard that "has no nationality" and "is eternally and universally accepted."

Two years later, de Gaulle again displayed his pique with the status quo, pulling France out of a system called the London Gold Pool.

The pool was a price-fixing cartel of eight governments: the United States, Britain, Germany, France, Italy, Switzerland, Belgium, and the Netherlands. It maintained the official gold price of the Bretton Woods arrangement: $35 an ounce. To stop speculators bidding up the price, the pool would dump gold into the market to satisfy any demand, always ready to sell at $35. But speculators, betting that this support could not last, amassed large holdings. They believed that the $35 price would crumble, and that it would have to rise as faith in the dollar fell. Not long after de Gaulle took France and its gold out of the pool, the United States had to transfer almost $1 billion in gold to London to sell at $35 to buyers who would otherwise have paid more, destroying the official price. The speculators saw the $35 price as a bargain. Gold, in this view, was on sale, and eventually that sale would end.

The pool members struggled against the inevitable. On March 11, 1968, only days after the huge bullion transfer from the United States, the pool vowed to hold the line at $35. Three days later, in a market that usually traded five tons a day, buyers in London snapped up a hundred tons of gold. A week later the pool folded. One year after that, the price of gold on the open market was $43 an ounce—a 20 percent premium on the official price. Something had to give.

WHEN RICHARD NIXON CAME TO OFFICE in 1970, money was pouring out of the country to pay for the Vietnam War and for the import of foreign goods that Americans increasingly preferred to their own, such as Japanese and German cars and electronics. Yet the main problem in the minds of American voters was the inflation that was driving higher prices. That was the problem Nixon had promised to solve. At the same time the administration had to face the other, potentially humiliating problem—what to do about the mountain of U.S. banknotes piling up in central banks around the world, each one of them an IOU for American gold. If rising prices caused by inflation disturbed most Americans, the worry of the banking world was the growth of dollar deposits outside the United States. For foreign governments, that was the pressing issue of the day. They knew they could not all show up at once to cash in dollars, because the gold demand would empty Fort Knox. The prospect of such an ignominy would probably cause the United States to head it off by closing the "gold window," and refusing to convert.

Moreover, the dollar anchored the global financial system. A run on the dollar would mean chaos everywhere. For these reasons the foreign governments held off. The tension grew. Anyone with large dollar holdings would necessarily worry about other countries with equally large dollar holdings. What if someone broke the tacit agreement keeping back the rush? What if one country reached a secret accord with Washington, and surreptitiously unloaded dollars while everyone else sat on theirs?

The solution to the American dilemma seems simple: Do just what the foreign dollar holders feared. Shut the window. Let the dollar float and find its value against other currencies according to what the market thought it was worth. Some of Nixon's advisers favored this course. But ending gold convertibility and letting the dollar float presented problems too. The dollar would drop in value. A lower dollar would mean that the imported products that Americans craved would be more expensive. It would look as if the administration had abandoned the struggle against inflation. In full view, Nixon would have failed in his pledge.

In the country at large, and to the world, the president would take the blame. In official Washington, though, attention fixed on the Treasury. The secretary was an outsized, dashing player whose appointment had raised eyebrows: a Democrat in a Republican administration. He was Nixon's antithesis—a bold, smooth-talking, swashbuckling lawyer who towered above the physically unimpressive president. He had been secretary of the navy and governor of Texas. His official portrait at the Treasury shows a man at the summit of his powers, sitting against a desk in a yellow room flooded with light, a commanding, confident, silver-haired man in a pin-striped suit, with one fist resting on his hip. This was John B. Connally. Nixon was besotted with him.

NIXON SURPRISED MANY WHEN HE appointed Connally to the Treasury in December 1970. Connally had been in President John F. Kennedy's cabinet, and Kennedy—attractive, privileged, and socially adroit—had inflicted a humiliating defeat on the awkward and uncomfortable Nixon in their televised debate in the 1960 presidential campaign. Connally had a particularly vivid association with the Kennedy mystique: as governor of Texas at the time, he had been riding in the open presidential limousine when Kennedy was shot, and had taken the same bullet that killed Kennedy.

The bullet that struck the president passed through him and hit Connally in the back. The images of Connally and the Kennedys, spattered in blood as Secret Service agents rushed to protect them, were seared into the world's collective memory. Connally's biographer believed that the assassination enhanced Connally's reputation. In a chapter called "The Bullet-Made Man," he said that when Kennedy died, Connally "was reborn. The same bullet that passed through

Kennedy made Connally a minor martyr, a role he was to exploit to such an extent that four years after the assassination the *Baltimore Sun* and other newspapers suggested with editorial sarcasm that 'the black arm sling that Governor Connally is so fond of wearing is getting frayed.' "

Connally was the perfect partner for Nixon. The president wanted an audacious solution to the crisis; Connally did not care what that solution was. Herbert Stein, who knew Connally and saw him in operation, and who wrote an account of the drama played out in Nixon's inner circle at the time, says Connally was "forceful, colorful, charming, an excellent speaker in small and large groups and political to his eyeballs." He brought no policy of his own to the Treasury. His talent was a knack for selling whatever policy the president might choose. "I can play it round or I can play it flat," he liked to say, "just tell me how to play it."

The administration's frontline challenge was inflation. Rising prices were eroding the living standards of Americans. One radical solution would be to cap price increases by law, and tie wage increases to the same restraints. Nixon's traditional Republican advisers were shy of such draconian interference with the free play of the economy. Connally had no such fear, and his department included technocrats who favored controls. It was natural in a man of Connally's emphatic nature to seek to establish his department in the forefront of the great issue of the day. In Stein's insider view, Nixon, who had tended to sample among the opinions of his circle, now formed with Connally a policy cartel of two. One of Nixon's closest aides, H. R. "Bob" Haldeman, the White House chief of staff, thought the president was in awe of Connally. The two met one-on-one. Stein concluded that Nixon isolated himself from more conservative advice, and probably wanted Connally to help him take a step he could not have taken by himself, or with the counsel of his usual advisers. The measures Connally and Nixon contemplated were political dynamite. They used the gold crisis to push the plunger.

In the spring of 1971, as overseas dollar deposits kept growing and the tension increased in foreign capitals whose central banks had more dollars than they wanted, Nixon and Connally made their decision. If any foreign dollar holder appeared at the gold window to convert, the Americans would close it. From that moment the dollar would cease to be convertible. At the same time, and most importantly from the point of view of the American voter, the government would impose wage and price controls. The policies would be linked, and sold to the public as a structural intervention on behalf of ordinary Americans, protecting them from both the crushing burden of rising prices and the predatory practices of foreign speculators. They would sell this so boldly, Nixon believed, that it would demolish the slightest chance of opponents saying he had not gone far enough.

Crucial to the success of the two-step plan was secrecy. If word leaked out, merchants and manufacturers would rush to raise prices in advance of controls. Foreign dollar owners would stampede the gold window. Instead of a decisive, forward-looking strike, the plan would crumble into a rearguard action fought to prevent the rout of a beleaguered economy. To preclude a leak, Nixon admitted only two of his most trusted advisers to the plan that he and Connally had formed. For months, only four people in the innermost circle of the White House knew that the United States planned to annihilate the system underpinning the world's finances. All Nixon needed was a pretext.

As the summer of 1971 went by, Americans focused on rising prices, and the rest of the world on the American dollar. In mid-

August, in a week of heavy dollar selling that pushed the value of the currency ever lower, Britain's *Guardian* newspaper reported that "whispers about the dollar's uncertain future have now turned into open forecasts of devaluation." The article went on to discuss the solution to the falling dollar favored by those who owned so many of them, particularly the French central bank: redeem them for gold. Then the writer stated what the whole banking world already knew: "There is not enough gold now in Fort Knox to meet more than a fraction of the total possible claims which could be made by foreign holders of dollars."

Two days later the *New York Times*, reporting "speculative attacks" on the dollar in Europe, revealed that, according to the Federal Reserve, "the nation's monetary gold stock had declined $200 million during the week [to] the lowest level since December 31, 1935." Adding to the systemic tension already in place (nervous foreign governments wanting to cash in their swollen dollar holdings but afraid of sparking a refusal to pay out) was the lethal reality of a two-tier price.

By mid-1971 the price paid on the open bullion market was $3 higher than the official rate of $35 an ounce set at Bretton Woods. There were, then, two prices for gold—one on the private London market and the "official" rate paid at the gold window in Washington. This discrepancy presented an irresistible opportunity to people with access to the window. They could buy an ounce of gold for $35 in Washington, and resell it on the private market for $38—an instant profit of 8.5 percent. The relatively slow traffic at the window was the result of the informal agreement among foreign government dollar holders not to menace the system by attempting to cash out of their massive positions. This détente slowed the flow of dollars into gold, but did not stop it. The United States continued to lose gold.

The easy profits harvested from the public-private price spread cycled back to the window to be traded in for yet more cheap gold.

For Connally and Nixon, the time to cut the country free from its golden shackles had arrived. The cumulative pressure was intolerable. International money markets accepted as inevitable that the Treasury would devalue the dollar. "Not only has the recent payments deficit of the United States been very large," the *New York Times* continued in the piece cited above, "but it comes as the culmination of an outpouring of dollars that stretches back, almost unbroken, to the 1950s." Obviously this imbalance had to be made good, and the world guessed that it would be. For the president's men, the chambers were fully loaded; they were just waiting for the right moment to pull the trigger. That moment arrived in the week of August 9. An envoy from the Bank of England came to the Treasury and asked for the redemption of $3 billion into gold. Nixon had planned exactly what to do. On Thursday, August 12, Connally cut short a Texas vacation and rushed back to Washington. The next day the president and his advisers helicoptered to Camp David to write the script that would topple gold from its place in the monetary system. Gold's final days as money had arrived.

Those who flew the sixty miles north to the presidential retreat on Maryland's Catoctin Mountain knew that they were making history. The participants included George Shultz, a future secretary of state, who then ran the Office of Management and Budget, and Paul Volcker, Connally's undersecretary for monetary affairs and a future chairman of the Federal Reserve. At the first meeting of the weekend, Nixon reminded everyone of the importance of secrecy. They were forbidden even to tell their wives where they were. Then Connally took over the meeting.

The drama of what they planned cloaked the proceedings in an air of high excitement. To Americans, the headline subject would be wage and price controls. But to people abroad, it was Nixon's other step that would stun them. It would change how the world worked.

The United States, although financially pressed, was still the economic master of the universe. The European Union was still only a six-country customs union. China's warp-speed rise as a business power lay far in the future. Russia's economy gasped along in a state of hypoxia. In that world, the United States could act as it liked in the monetary sphere. It did not have to cajole allies or build coalitions. It was prey, like other nations, to the costs of war and the price of oil and the penalties of liking what other people made more than what it made itself. But its economy dwarfed all others and its leader was always called, at least at home, the world's most powerful man. For those around the president that weekend at Camp David, the sense of this power, at a moment of great consequence, charged the days with an exhilaration that seems to have had, as Stein recorded, as much to do with the participants' sense of privilege as with the momentous acts afoot.

The retreat had had a storied place in American affairs since President Franklin D. Roosevelt started using it in 1942, calling it Shangri-La and modeling the main lodge on the Roosevelt family winter home at Warm Springs, Georgia. In 1953, President Dwight D. Eisenhower renamed the hideaway Camp David, for his grandson. Idyllic cottages named after trees are tucked into a forest of oak, poplar, ash, locust, hickory, and maple. A high-security fence keeps trespassers away from the 125-acre sanctuary. Administered by the White House Military Office and staffed by Navy personnel, the retreat gave its guests a sense of privilege and power. Although the accommodations were simple rather than luxurious, "every provision was made for the wishes of the participants," Stein recalled, including "any choice of food and drink, tennis, swimming, skeet shooting, bicycle riding, horseback riding." The Navy staff that ran Camp David "were unfailingly helpful and courteous, treating everyone as if he were a full admiral."

Coddled, guarded, and cut off from communications below, the president's men settled down to the task at hand—drafting the bombshell that Nixon would explode on live TV.

The news played into the administration's hands. As the president's men had chopped out of the capital to their retreat, the *New York Times* was warning its readers that international financial arrangements were unraveling fast. Calling the situation a "ferment," the paper pointed out that "three major currencies, the German mark, the Canadian dollar, and the Dutch guilder [were] all 'floating' without reference to a fixed par value—strongly suggest[ing] that the system of fixed foreign exchange rates that the non-Communist world has used since World War II may be close to its end."

Word of the top-level meeting at Camp David leaked out, and heaped more logs onto the fire of speculation crackling through foreign capitals. Britain's *Observer* reported that Nixon had closeted himself with his advisers to deal with a crisis that was "becoming deeper day by day" and that had caused a "wage-price spiral which is affecting every American family." More to the point for British readers, whose country had large dollar holdings and whose representative was in Washington at that moment trying to cash them in for gold, the newspaper quoted the U.S. Treasury as stating: "We are dedicated to preserving the integrity of the dollar. We are not going to devalue it or change the price of gold. Our position on this is unwavering."

This was strictly true. They were not going to devalue the dollar or change the gold price. They were going to unhook them.

Most of the Camp David planners were exhilarated by the weekend, but Nixon fretted about timing. On the brink of dismantling the global financial system, he worried about interrupting the hit TV western *Bonanza*. Millions of Americans sat glued to their televisions every Sunday night, addicted to the fictional Cartwright family, whose adventures played out on the Ponderosa ranch. Nixon was not the only president to respect the devotion of *Bonanza* viewers. President Lyndon Johnson had made it policy never to let the affairs of state disrupt proceedings at the Ponderosa. But Nixon's advisers convinced him that he had to speak before the markets opened Monday morning.

Bob Haldeman, visited the president in his private cabin at Camp David on the Saturday night, one day before his epochal address. "The P. was down in his study with the lights off and the fire going in the fireplace, even though it was a hot night out," Haldeman recorded in his diary. "He was in one of his sort of mystic moods." Nixon told Haldeman that "we need to raise the spirit of the country; that will be the thrust of the rhetoric of the speech. . . . We've got to change the spirit, and then the economy could take off like hell."

On Sunday, as the president's political and technical advisers worked to finalize the announcement, the news must have stiffened their resolve. In a piece from Brussels datelined that day, the *Times* described a scene in which the European market was shunning dollar bonds. One dealer characterized the trading as more like a flea market than a money market. The price of dollar bonds was dropping faster than computers could track. The number of dealers ready to trade dollar instruments at all had dwindled to a handful. Ready for their moment, even exhilarated, the presidential party returned to the capital, and that night, on television, the president made the announcement that rippled out through an astonished world and was known ever after as the "Nixon Shock."

> *In the past seven years, there has been an average of one international monetary crisis every year. Now who gains from these crises? Not the workingman; not the investor; not the real producers of wealth. The gainers are the international money speculators. Because they thrive on crises, they help to create them.*
>
> *In recent weeks, the speculators have been waging an all-out war on the American dollar. The strength of a nation's currency is based on the strength of that nation's economy—and the American economy is by far the strongest in the world. Accordingly, I have directed the Secretary of the Treasury to take the action necessary to defend the dollar against the speculators.*
>
> *I have directed Secretary Connally to suspend temporarily the convertibility of the dollar into gold or other reserve assets, except in amounts and conditions determined to be in the interest of monetary stability and in the best interests of the United States.*

The first market with a chance to react was Tokyo, opening for business as the president's words flew across the Pacific and detonated on the trading floor. In a single hour, Japan's commercial banks unloaded $300 million in U.S. cash. At the same time Japanese investors started dumping the stock of companies with big American export markets, for Nixon had done much more than slash the mooring line between the dollar and gold. He had announced a package that included an import duty on foreign manufacturers. As dismay spread around the globe, the White House awaited the American response. They needn't have worried. It was jubilant.

And why not? Prices would stop rising. The tax on imports was sugarcoated by the cancellation of a federal tax on buying an American car. Money straight into the voters' pockets, and good for the country too! Everything announced that night was bathed in a patriotic light. Nixon had worried that closing the gold window would signal Americans that their government was bankrupt, so he played it as a defense of the home currency against foreign speculators. When the TV lights clicked off in the Oval Office at the end of the address, the United States dollar had changed from a drawing right on gold to a paper note whose value would thereafter be a matter of opinion. But what had gold become?

Gold metamorphosed. It shape-changed in the human mind. No longer hard currency, or even its relation, gold became the phantom money of the imagination—shadow money. The shapes projected were the shapes of our emotions. "When you wake up in the morning, do you care about the price of gold?" the Texas gold bug Ron Paul asked Ben Bernanke, chairman of the Federal Reserve, four decades after the Nixon Shock.

"I pay attention to the price of gold," Bernanke said, "but I think it reflects a lot of things. It reflects global uncertainties. I think the reason people hold gold is as a protection against what we call tail risk—really, really bad outcomes."

One of the greatest goldfields in the world, Nevada's Carlin Trend, was discovered when gold was still $35 an ounce, and then rediscovered when the price had been cut loose. The story of those discoveries in the low mountains of northern Nevada describes the transition from the old gold world to the new. On the Carlin Trend a great tycoon built the biggest gold mining company in the world. Yet the story didn't start with him. It started in another age, when time moved at a more dignified pace. It unfolded among the piñon groves and the empty hills with a kind of purity of purpose, moved less by material reward than by the desire for scientific truth. It uncovered an ocean of gold.

5

THE DISCOVERY OF
INVISIBLE GOLD

While I was working in the field, cresting the ridge was always
important to me, for then I could take a breather, and look ahead
across the valley. —Ralph Roberts, *A Passion for Gold*

RALPH ROBERTS FOUND THE RABBIT HOLE INTO
Wonderland. The greatest goldfield in America lay hidden under a
thick cap of rock, and Roberts found a place where a hole in the cap
allowed him to look through. By the time he made his discovery,
he knew the mountains well. He had wandered for years through
the Tuscarora and Shoshone, the Sonoma and the Diamond and the
Antler ranges. They kindled his excitement. To a geologist, the fea-
tures of a landscape are like jumbled parts of speech, to be construed
into the sentences of geologic time. It is all just heaped-up rocks and
sand until it is fitted into a comprehensive theory, the story of what
put it there.

The reader finds in Roberts's account of his life the deep pleasure
of inquiry, and also of freedom. "On those trips into the mountains I

was accompanied only by a few sparrows, meadowlarks, and shrikes, who actually flew alongside me when I was near their nest." He saw mule deer and antelope and golden eagles. "Marmots announced my approach with piercing whistles, and voles darted from underfoot. I seldom saw another person, except at ranches in the foothills, and that was good because I didn't want company while learning the secrets of the rocks."

To walk into the hills that Roberts explored is to enter an enormous, sealike silence of mauve and ocher slopes. The ground is speckled with the tough little plants of the western desert. In the spare hills, the planetary history opens to inspection.

ROBERTS WAS A TWENTY-EIGHT-YEAR-OLD DOCTORAL candidate at Yale University when he joined a United States Geological Survey exploration team in north-central Nevada in 1939. He stepped off the bus in Winnemucca into the furnace of July, and within days was puzzling over the events of half a billion years ago. The expedition's task was to untangle the geology of the Sonoma range. As he tackled the exploration, Roberts caught a lucky break. There was a working gold mine in the heart of Willow Creek, where he was based. The mine would provide him with a window into the structure he was trying to unravel.

Mining in Nevada had a long pedigree. About 13,000 years ago, Indians of the Clovis culture mined deposits of obsidian, opalite, chalcedony, agate, jasper, and quartz. They made spear points and arrowheads and tools for scraping hides. From about AD 300 to 1200 the Anasazi people mined turquoise near present-day Boulder City. Spanish explorers may have crossed the southern tip of the state in 1776 looking for gold, silver, and turquoise. Native Americans probably knew about gold in the Sierra. There are reports that they showed whites where to look.

The first discovery on record came in 1849, when a party of Mormons heading for the California gold rush camped at Carson River to wait for the snow to melt from the mountain passes. They panned the streams, and recovered gold. They moved on westward in the spring, but Gold Canyon, as the site became known, attracted other placer miners. By 1857 successive waves of prospectors had pushed upstream, always finding gold. A second discovery, to the north in Six-Mile Canyon, brought more searchers into the foothills. Finally in 1859 they hit on where the gold was coming from—a rich gold-and-silver deposit near Virginia City, the famous Comstock Lode. Soon the passes were swarming with men leaving California for the new bonanza. Prospectors poured into the state and fanned out through the mountains. They found placer gold in the streams near Battle Mountain and Winnemucca. In the clefts and watercourses of the hills that Roberts was exploring, you can still see sluices and stone dams and the collapsed entrances of adits. The gold mine at Willow Creek that Ralph Roberts wanted to explore dated from that teeming gold camp.

The mine belonged to Wallace Calder, a Winnemucca dentist. Calder told Roberts he'd recovered large nuggets from a fault zone in the mine, and Roberts arranged a visit. Calder showed him where the nuggets came from. Examining the rocks close up, Roberts saw that they were oceanic rocks, much older than the rocks they sat on. In the normal course of geological formation, new rock comes up from the mantle in molten form, and is deposited on top of older rock. Here, those relative positions were reversed. To explain what had happened, Roberts hypothesized a thick shelf of older oceanic rocks thrusting up onto the younger limestone at what had been the western edge of the continent. By the end of the summer, although he was years away from forming a theory about the region's gold, two pieces of that theory were now in place in his mind: first, older ocean rocks had been pushed up onto younger rocks at the ancient seashore; and second, that gold occurred within and below the zone where that thrust occurred.

Today, geologists recognize the thrust as the product of a tectonic collision, where two of the earth's crustal plates mashed against each other in the geologic past. But the plate theory did not emerge until the 1950s. In 1939, when Roberts began his explorations, the most important guide to the regional geology was a nineteenth-century report called the Fortieth Parallel Survey. Commissioned by the secretary of war and published in 1877 and 1878, the survey was an eight-volume natural history of parts of Nevada, Idaho, and Wyoming. The writers described huge tables of Triassic rock and mountainsides alive with "minute brilliant-black crystals of tourmaline" and "small brown iron garnets, not much bigger than a pinhead." Field geologists have an unquenchable appetite for such details. Diamond hunters obsess over garnets, which can point the way to diamond pipes. Roberts's interest was the larger structures of the region, and one day the camp cook dropped him off at an abandoned silver mine. He struck out along the Antler range in his happiest state—alone.

Earlier geologists had identified the range that Roberts followed as 300-million-year-old limestone. Limestone is composed of the skeletal remains of tiny animals from the shallow seas, in this case, seas that had once lapped at the continental shore in present-day Nevada. The limestone takes its pale color from the minute skeletons that make it up. When the ocean plate and continental plate collided, according to Dean Heitt, a Newmont Mining Corporation geologist who has written a history of the Carlin Trend, "the darker, older rocks from the deep ocean, such as cherts and shales, pushed up onto the younger, light-gray limestone, covering it." When Roberts set out to understand the regional geology, finding a theory to explain how the older rocks had ended up on top of the younger rocks was part of the exercise. "As I approached the western margin of the range," he wrote, the "limestone pinched out," and older rocks appeared on the ridge he was traversing. "I could see a few fragments of limestone on the ridge, but the massive unit that I had followed earlier was gone."

Roberts did not know it, but he had crossed what he would come to call a "window"—a place where the older, darker rock had worn away and revealed the younger limestone underneath. At that break in the older rocks that had been thrust up from the deep ocean, Roberts located an opening into the richest gold-bearing rocks in America. He was "seeing through" the cap of older rock that concealed the limestone. In the Willow Creek mine he had already seen a hint of what he would later understand more fully—that gold lay in the limestone layer below the older rocks. He had come across a place where an explorer could reach in and rummage for gold without having to penetrate the barrier of older, harder rock. That Roberts did not see this shortcut right away is because he hadn't started looking for it.

Roberts's work was interrupted by World War II. He joined the wartime search for strategic minerals, and was posted to Central America. He did not return to Nevada until 1954, when he took over a unit mapping the geology of Eureka County. That appointment set him on the path to his discovery.

In Roberts's autobiography, written when he was ninety-one, an unselfconscious brio lightens the pages. The reader cannot help but feel that here was Fortune's darling, a man delighted by the life he found in front of him.

RALPH JACKSON ROBERTS WAS BORN in 1911 in Rosalia, Washington. Both parents came from nearby wheat farming families. As Roberts tells it, his father got tired of slinging 120-pound sacks of grain, and left the farm to become a druggist. He tired of that too, when the long hours of keeping a small-town pharmacy wore him down, and he took to selling Edison record players. Finally he bought a candy store in Omak, eastern Washington, a town in the Okanogan Valley. While he made "incomparable peanut brittle, nougats, pinoche, divinity, caramels, butterscotch, taffy, and fudge," his son took over the soda fountain.

For a future geologist, Roberts was lucky with his teachers. One was a first-rate chemist and another ran a thulite mine. Thulite—sometimes called rosaline for its blazing pink color—is a crystalline mineral cut into slabs for use as decorative facing. Roberts helped to quarry it. He also made forays into the surrounding hills to look for gold with "Dad" Hayes, a handyman who had lost the fortune he'd panned from Alaskan gold creeks.

Roberts picked up mining lore from his family too. One of his mother's uncles had a silver mine in Colorado, and uncles and cousins on his father's side had filled him with tales of their mining exploits. Even the Roberts Mountains of north-central Nevada, his exploration ground, had been named for a distant relation.

Roberts's account hurtles along. Dancing lessons in wartime Washington occupy the same chapter as the development of the cordilleran geosyncline 380 million years ago. In the same six pages he meets his future wife, a classmate's girlfriend. "Our eyes met and held for a long moment. She was wearing Warren's ring, but her greeting to him seemed restrained. . . . I do not remember much about that visit, except that I was drawn to Arleda. Since she was Warren's girl, I tried not to let my feelings show, but I suspect that she knew I found her attractive, and I sensed that she was interested in me."

He sensed right. Two years later Arleda wrote him out of the blue to say that she had returned her engagement ring and would like to see Roberts again. He took her to dinner and proposed before they reached dessert. They had a long, eventful, happy marriage. A photograph from 1954 shows her smiling in the sun with their three children in front of the forty-two-foot trailer where they lived when Roberts set out to map the geology of Eureka County, and made his great discovery.

Roberts mapped northward through the Cortez, Shoshone, and Tuscarora ranges. He saw that the feature called the Roberts Mountains thrust presented a continuous geological relationship—deep-ocean rocks pushed up over shallow-water limestone. He believed that the thrust was a regional feature, not a local one. He had uncovered the basic geology of what is now called the Carlin Trend. He had taken a giant step toward unmasking the deposit, although another crucial step remained: realizing it was there. Certainly there was historic evidence of gold from early placer mining, and the Willow Creek mine had yielded nuggets. But such occurrences don't by themselves suggest a single massive source. Yet as Roberts was coming to understand the geology of the thrust, another man was scrutinizing local gold production, and realizing that some of the gold was coming from ore unlike any other he had seen. One characteristic distinguished it. The gold was invisible.

William Vanderburg, an engineer with the U.S. Bureau of Mines, had been making reconnaissance surveys of Nevada since the 1930s. He had regularly visited the mines and described deposits. He had been struck by the observation that at some gold mines the particles were so fine they could not be recovered by panning. The mines were recovering gold, but when Vanderburg panned a sample of the ore he got nothing. It was "impossible to distinguish between ore and waste except by assay," he wrote, "and gold is present in such a state that it is impossible to obtain a single color [visible gold] by panning." At one point there must have been visible gold, or there would have been no mining. At some point, then, the miners had used up that original reserve of ore, but found that the recovery mill was still producing gold even when they moved beyond the original deposit and started feeding in rock that did not have visible gold. They kept getting gold, so they kept mining.

"Bill wanted to show me this unusual ore," Roberts recalled. "My curiosity was piqued, so I went with him." The manager of the mine they visited, Gold Acres, led Roberts and Vanderburg to the "pay dirt"—the gold-bearing zone. Roberts saw the familiar zone of shearing, where the older ocean rocks had pushed up over the younger limestones at the continental edge. "For anyone who sees this contact zone in the field," Roberts wrote, "the color contrast between the dark siliceous oceanic rocks and the underlying gray continental shelf carbonate rocks is striking!"

Unlike the narrow shear zone that Roberts had seen in the Willow Creek mine when he had first visited Nevada as a student, the shear zone at Gold Acres was in places more than a hundred feet thick. Here and there throughout the zone were limestone "lenses"—slabs of limestone torn from the continental rock when the oceanic rock had scraped across it. These lenses contained the ore.

To Roberts, this was a thunderous discovery. If the gold was contained within the shear zone, and the shear zone was a regional feature—*the* geological feature of that part of the state—then the gold occurrences were not local accidents, but part of a structure that ran through the whole region.

"When we entered Maggie Creek Canyon," Roberts wrote, "I saw for the first time the stunning view of upper-plate black chert riding on iron-stained gray limestone on the east side of the canyon. . . . This visit was a profound and moving experience for me as it confirmed that the thrust was of regional extent and might exert a regional control of mineralization."

By "control," Roberts meant that the thrust might be the feature that determined where the gold was to be found. In Maggie Creek he saw that zone exposed to view. It was in limestone that the invisible gold was to be found. The limestone was more porous than the older rocks that sat on top. Gold is formed in the mantle, and rises in a hot solution. If the gold had flowed up into the porous limestone at a point in time after the harder, denser older rocks had thrust on top, the older rocks, Roberts thought, might have "capped" the gold, blocking its flow and penning it in the limestone.

Sometimes when gold penetrates rock, it settles into faults and cracks, cooling into solid veins. This type of deposit is called a lode. But in the deposits Roberts envisaged, an acidic gold-bearing solution had flowed into the porous limestone, dissolving the rock and creating pathways for more gold to follow. The picture that formed in Roberts's mind was of a gold-rich layer of limestone capped by harder, less permeable rock. The harder rock had trapped the gold inside the limestone. "My excitement grew as I visualized potentially rich ore bodies," Roberts wrote.

Roberts does not say if he discussed his ideas with Vanderburg, but according to Alan Coope, a geologist who wrote a history of the Carlin exploration, Vanderburg too had been seized by the conviction that more gold awaited discovery. He thought that explorers would find gold in the kind of shale-and-limestone sandwich that Roberts was identifying as the defining character of the region. Roberts saw the bigger picture. He saw past the gold mine to the goldfield. He knew that where gold had been discovered, the upper layer of rock had worn away, leaving a window into the softer rock below. As he saw it, the gold mines of the region were not located on separate areas of gold-rich ground, but on the same vast zone, one that had been tapped, you could say, by accident. Miners had blundered onto deposits through windows opened by erosion, without understanding the larger structural picture.

Roberts's mission was to map the structures of Eureka County. He led an idyllic life, camped with his family in a trailer on the Dean Ranch in Crescent Valley. His sons explored nearby ghost towns while Roberts and his assistants traced the thrust zone through the hills. Increasingly convinced that the thrust was a main control of regional gold deposits, and looking for a pattern to confirm this conjecture, Roberts made a rough map that plotted the ore deposits and the windows on the thrust. A clear line jumped out. It trended northwest through Eureka County. In 1960 Roberts put his thoughts into a two-and-a-half-page paper, "Alinement of Mining Districts in North-Central Nevada." He asserted his belief that windows in the cap of older rock gave access to zones that had been "penetrated by conduits along which igneous rocks and related ore-bearing fluids rose. The zones probably penetrate to great depths within the crust. . . . In prospecting within the windows, a special effort should be made to explore the lower units." In other words: Dig the windows, and dig deep.

Roberts perceived a condition that he could not see. Like an astronomer deducing the existence of a body outside the range of observation, he created a hypothesis out of what he knew. He could see that microscopic gold was present in pieces of limestone. He could see that the limestone was part of a regional structure. From these observables he built the mental image of a goldfield of invisible particles trapped beneath the older shale. The only way to test that theory was to drill the exposed limestone in the windows. As a government geologist, it wasn't Roberts's job to drill, but to convince someone else to drill.

The man Roberts found was an intrepid loner—like Roberts, a man who loved the silence of the mountains. His origins were very different from Roberts's: born into one of America's grandest families, he had left behind a privileged existence at the top of the social world to pursue a solitary life. For years he and Roberts, unknown to each other, had explored the same mountains at the same time, in search of the same gold.

Roberts's paper on the windows appeared in 1960. The next year he addressed a meeting of geologists at a hotel in Ely, not far from Eureka. Most of those present were oil geologists, not much interested in Roberts. His talk came at the end of the day. As soon as it was over, the room emptied out in the direction of the bar. One man stayed behind.

John Livermore was a lanky, six-foot-five-inch gold geologist who had read Roberts's paper. It had stunned him. On and off through the 1950s, he had explored the mountains searching for exactly the kind of gold deposit Roberts had described, and in that short paper on the alignment of mines, he read where to look.

• ● •

JOHN SEALY LIVERMORE GREW UP in San Francisco, in a redwood mansion on San Francisco's Russian Hill. His name and those of his brothers are preserved in a stained-glass window above the south transept of the city's Grace Cathedral. Mount Livermore, on Angel Island in San Francisco Bay, was named for Livermore's mother, Caroline, a conservationist who helped preserve it. The family spent summers at Montesol, the 7,000-acre ranch in the hills north of Napa that the Livermores had owned since 1880. Livermore admitted to a "pretty perfect childhood," but the life it pointed to, of social prominence, did not attract him. With a geology degree from Stanford University, he took up the nomad's calling and became a prospector.

Livermore's search for invisible gold began in 1949 at the Standard mine, a gold mine north of Lovelock, Nevada. It had closed during World War II. The owners wanted to start it up again, but needed to find a new supply of ore. The mine had been established on a lode deposit. If the host rock was fractured, Livermore thought, or if there had been repeated gold intrusions, particles of gold might have been forced outward into the surrounding rock in a kind of low-grade mist of gold. Livermore found plenty of such low-grade rock, but not the richer, higher-grading source rock that should also have been there.

"We did some churn drilling," Livermore later recalled, referring to the use of wide-diameter drills, "and found some ore." But the ore was not high-grade, and "the mine was really sort of doomed already. It was a very, very low-grade mine, and they were not making much money. To develop this ore that we were finding would have involved quite a lot of stripping [removing non-ore-bearing rock], and they just weren't generating enough money to continue." In a few months the mine shut down and Livermore was out of work. He tried to find another job, and when he couldn't, did what he liked best—trekked out into the land and looked at rocks.

For two years he prospected alone in the Nevada mountains. He had been hooked on the idea of deposits of undiscovered gold in that part of the state since reading a paper by William Vanderburg, the engineer who had taken Roberts to Gold Acres. As Livermore searched, he wondered if the low-grade ore of the Standard mine might represent a different, unidentified kind of gold deposit, and not a low-grade example of an ordinary lode deposit. To fit the accepted model of a lode, the gold should have occurred in hard, volcanic rocks. Instead, it was in softer rocks, such as limestone. Moreover, the grains were so tiny as to be invisible. Only an assay could detect their presence. But if this invisible gold was a new kind of deposit, what accounted for it? Livermore needed to imagine some new process that would enable such mineralization to take place, a theory to explain the presence of the gold. He set about to build one.

In Livermore's conjecture, the gold rose from the mantle into the crust in the same kind of solution that created normal lode deposits. But instead of the gold precipitating out into the wall rocks of the channel and concentrating there, it flowed into relatively soft and porous limestone, dissolving some of the rock to create yet more spaces to flow into. In north-central Nevada the rocks were already cracked and broken from seismic action, providing more routes into the limestone for the hot, acidic gold-bearing solutions that Livermore conceived. He was imagining what Roberts later would hypothesize.

In 1952 Livermore took a job with Newmont Mining Corporation. For the next six years he worked on projects in South America, Morocco, and Turkey. In 1958, recovering from an illness, he took a desk job at the company's headquarters, then in New York. He found it tedious, and his thoughts returned to Nevada. In 1960, on assignment at a base metals mine in Eureka, he persuaded his boss to give him time to look for gold.

Livermore has described the warring emotions that accompanied him through the empty landscape. "I really and truly believed I would find gold." Yet this feeling was tempered by doubt. As a geologist he knew that the chances of discovering an economic gold deposit were slim. "Gold will always surprise you—more than any other mineral, I think," he told *The New Yorker*'s John Seabrook almost thirty years later. "You can't predict it. You get some decent gold values in a sample and you sort of think to yourself, Aha, I'm onto you now, I've got your secret this time. Then it turns out you don't—the gold was just playing with you."

Livermore persisted anyway, consoled by his affection for the land, even as it bruised his hopes. "I get lonely sometimes, too," he admitted, "living in those motel rooms, spending all day by myself. I ask myself why I'm prospecting. It's not to be rich. It's nice to have the money, but, honestly, that's not why I do this. I don't know. It's something about—about the finding. If I could just find that gold, then everything would be OK. It's this endless puzzle, and sometimes—I don't know—it seems more important than it really is."

It wasn't even worth that much—$35 an ounce. The only large gold mine in the United States at the time was the Homestake mine in Lead, South Dakota—at 8,000 feet, the deepest mine in America when it closed in 2002. Homestake produced 40 million ounces in its lifetime, and a belief had taken root among miners that there were no great deposits left to be found in the United States. And if there were, they would not be found in north-central Nevada, scraped and sifted for a hundred years. Yet that is where Livermore found it, not so much a deposit as a sea of gold.

When he saw Roberts's paper, with its simple plot of existing mines, mines already harvesting gold through the windows, "That really got me interested," Livermore said, "because now I had a model to follow to try to find some of these deposits. I thought there was a good chance they existed. Maybe the old-timers would have missed them because of this fine gold, but now with his theory, maybe this was a way to look for these deposits. Maybe just by carefully mapping and prospecting this Roberts Mountains thrust, we could find more of these deposits."

With nothing fancier than hammers and canvas sample bags, Livermore and an assistant set out along the thrust. They sent their samples to Harry Treweek, an assayer who lived with his wife in a secluded cabin in Crescent Valley. Assayers heat ore with chemicals called fluxes. The fluxes combine with everything but the gold, which settles out of the mixture, where it can be weighed. Its weight is expressed as fraction of the tested ore, in ounces or grams per ton—the grade.

"We had confidence in him," Livermore said of the assayer he used, "because these samples we were taking were often very low-grade samples, and we wanted to be sure that what he was reporting were true values, and not just spurious values."

Livermore's discovery can look simple: Roberts tells him where to look and the rest is easy. But the windows that provided openings into the prospective limestone were scattered over 5,000 square miles. In Livermore's estimate, he mapped a mile a day. For two men based at a motel in Carlin, armed with nothing but rock hammers and a map, the exploration was a daunting task. Then Livermore visited Harry Bishop, manager of the Gold Acres mine, where Roberts had first examined the ore that contained invisible gold. Livermore asked Bishop where he would look himself, if he had the whole Roberts Mountains thrust to search. Bishop told them he would start just north of Carlin, where explorers had found gold at the Blue Star turquoise mine.

In June of 1961 Livermore started his examination of the Blue

Star mine. He mapped and sampled the deposit for three weeks, sending rock to Harry Treweek in Crescent Valley. When his assessment was complete, Livermore had outlined 500,000 tons of ore. He recommended the property to Newmont. When Newmont couldn't close a deal with the owners, Livermore moved to another target, and it wasn't far away.

In his weeks at Blue Star, Livermore had come to understand the formation more intimately. In the model he developed, upward-migrating hydrothermal fluids, loaded with gold, had "ponded" beneath a particular structure of the thrust, concentrating large amounts of fine-grained gold. Using his new knowledge, Livermore started mapping in an area now called the Lynn window, near the Blue Star mine. He sampled the rock and the assay "kicked"—showed gold. In September Livermore staked seventeen 20-acre claims, buying the posts and driving them in himself. Newmont hired a bulldozer, and in October started trenching through the rock near a peak in the Tuscarora range. One of the trenches intersected eighty feet of rock grading .2 ounces of gold per ton. In November a heavy snowfall ended the exploring season, but they were back in April with a drill. The third hole intersected an eighty-five-foot-thick band of ore. "It was that same submicroscopic gold I had first seen at the Standard mine," Livermore said, "uniformly disseminated, no major metallurgical problems—it was just a beautiful ore body. My God, it was a beautiful ore body."

Livermore made the strike on a height called Popovich Hill, named for a prospector who had kept a cabin there. Popovich had lived on the mountaintop for years. All around him lay the Tuscaroras. I suppose he tramped through every inch of them. In John Huston's classic gold hunt movie, *The Treasure of the Sierra Madre*, the old prospector says: "I know what gold does to men's souls."

Popovich left no trace except his name and an abandoned cabin. We don't know a thing about him except where he slept his hopes away, on top of the biggest goldfield in America—invisible. In two years of tearing up the peak where Popovich had lived, Newmont outlined a reserve of 12 million tons.

An ocean of new gold lay ready to be tapped, and yet . . . there was no gold rush. Livermore had identified a previously unknown category of commercial gold deposit on a trend that struck for forty miles. It was five miles wide and sometimes hundreds of feet thick. Yet only one other significant discovery was made in the 1960s— at Cortez, in a valley south of Carlin. Livermore was dispatched to Canada. Other than a single Newmont mine that opened in 1965, a goldfield that now has mines strung along it like Christmas lights sat there largely unexploited. The reason was simple. The gold price was too low.

6

GOLDSTRIKE!

He was the greatest gold miner of the modern age—
a silvery, immaculate, dashing, and indefatigable tycoon.

NO ONE FOUND MORE MINES IN NEVADA BECAUSE they were not there to find. The gold price hadn't put the ore in place. Ore is a human construct, not a natural one. Nature makes metal. Ore is made by math. The definition of ore is rock that can be mined at a profit. Ore comes into existence only when the value of a mineral exceeds the cost of getting it out of the rock. When Livermore made his discovery the price of gold was $35 an ounce. To get that $35, Newmont had to build a mill, strip away the overburden, blast and dig up the rock, truck it to the mill, smash it into a powder, feed it through a complex of special tanks in which the particles of gold attached themselves to carbon, collect the mess at the end, burn off the dross, and pour it into bars. All this consumed capital, energy, and wages. There was only one place for Newmont to recover those expenses—from the $35, ounce by ounce. In 1965, when the Carlin mine got going, enormous deposits of gold-bearing rock lay all through the hills, but they did not qualify as ore because they were not rich enough to support mines. The Nixon Shock changed that.

When the gold convertibility of the dollar ended in 1971, the true price of gold—what private buyers would pay—had been rising. Now that rise gained speed. Three years later the gold price passed $180. Large swaths of north-central Nevada turned into ore. A second phase of mine development swept the Carlin Trend. Then the party really got going.

In the last half of the 1970s a series of crises—the Islamic revolution in Iran, the rising oil price, an American recession—caused some investors to buy bullion as a hedge against jittery currencies. The gold price rose. Then in a single year, from 1979 to 1980, it shot into the stratosphere, blasting from $222 an ounce to $825. The scale and explosiveness of this increase was unprecedented. A firestorm of hype and speculation crackled around it. Gold captured a new place in the public imagination. It became what analysts call a "sentiment indicator"—the measure of emotions such as fear, greed, or panic. Gold fever seized the market and common sense went home. "Nobody in his right mind will sell at these levels," a New York dealer told the Associated Press in December 1979, when the price passed $500. "They're afraid they'll never be able to buy it back."

But they could have. In two years the gold price fell to $290. Then it recovered, and by the 1990s had settled into a range between $300 and $400 an ounce. As an investment, gold would never be the same again. A new scheme for rating gold mine stocks (in which the upside of possible price rises was incorporated into the structure of valuing the stocks) became the industry standard. In less than ten years the stock market value of North American gold miners increased 150 times, from $200 million to almost $30 billion, as millions of ounces of gold poured out of north-central Nevada. The fortunes of the two biggest gold mining companies in the world were founded on the Carlin Trend, one of them by a man who had learned, through bitter experience, to fear the gyrations of commodity prices, and who found a way to flatten them. He was the greatest gold miner of the modern age, a silvery, immaculate, dashing, and indefatigable tycoon with the menacing aplomb of a Florentine prince.

PETER MUNK WAS BORN IN 1927, heir to a Hungarian business fortune. In 1944, when the German wartime occupying force started rounding up Hungary's Jews for shipment to the death camps, the sixteen-year-old Munk and his family were among those rich enough to buy their way out on the Kasztner train, a transport arranged by Rezso Kasztner, a Jewish lawyer, in exchange for the payment of a bribe in cash, gold, and diamonds to the death camp planner Adolf Eichmann. Munk reached safety in Switzerland, and later emigrated to Canada.

Escaping a degrading death in one of history's monstrous passages must mark a person. Munk does not like to talk about it, or anyway, not to me. An account of the events already exists in Anna Porter's *Kasztner's Train*. Porter, born in Hungary herself, is a Toronto publisher, and friend of Munk's. She described sitting with Munk in his house, in a room darkened by heavy drapes, with Persian carpets on the floor and, on the wall, a picture of Munk's grandfather's house in a wealthy Budapest quarter, with the old man standing in the garden. "I've been sorting some boxes," Munk told her. "I don't think about the past much, but we have to pack some of these old things."

Munk's parents divorced when he was six. At first he lived with his mother. Later, when his father remarried, he shifted back to his Munk grandfather's house. When the German army entered Budapest in March of 1944, Munk's routine had him traveling back and forth across the Danube on visits to his mother, Katharina. This trip soon became more difficult. Within days the Nazis published edicts requiring Jews to wear yellow armbands displaying the Star of David, and forbidding them to use public transit. Munk kept up the visits anyway. He had fair skin and blue eyes, and did not look obviously Jewish. He stuffed the yellow armband in his pocket and hopped the tram across the river, risking his life.

The deportations started, and in early June of 1944 Munk's paternal grandfather, Gabriel, bought fourteen tickets on the Kasztner train, paying for them, Munk recalled, with a suitcase stuffed with gold and cash. From the young man's point of view, the family escape plan had one chilling drawback: it did not include his mother.

Gabriel felt no responsibility to include Katharina, since she had left the family. Munk told Anna Porter that for the first time in his life, he stood up to his grandfather, declaring that he would remain behind in Budapest with his mother. Katharina broke the deadlock herself, insisting to her son that he accompany his father's family. To his mother's wishes, Munk's father added that Katharina would only be temporarily detained. "My father assured me that if we left enough money for food and bribes, my mother would be fine," he said. "But I think he knew she would be deported."

In a haunting scene in Porter's book, Munk recalls escorting Katharina to the detention center in a former rabbinical seminary. She wore a cream-colored summer suit and Italian shoes. He carried her elegant leather suitcase. She had her money in a small purse. They talked about what they would do together when the war was over. They said good-bye at the gate. The next day, about to leave on the Kasztner train, Munk called his mother at the detention center. Amazingly, they put him through. Katharina said she was fine, and that her maid had brought fresh laundry. Munk boarded the train for Switzerland. A few days later the Germans put Katharina in a boxcar bound for Auschwitz.

Katharina survived, but she emerged from Auschwitz painfully altered. She returned to Hungary. As Porter told me when I called to talk to her about this part of Munk's story, he "wanted desperately to be with her, after discovering that she had survived, and *what* she had survived. He adored his mother."

Katharina insisted that Munk go to Canada to complete his education. Later he brought her there, although she never liked the country and did not speak English. "He took very good care of her," said Porter. "That was the strongest relationship in his life."

When I met Munk, he was eighty-four years old, a graceful, tough-looking, slender man with a hawkish face. Short white hair curled tightly on his head. He moved with an athletic step. We met in his fifth-floor corner suite at Claridge's Hotel in London. An unseasonably warm spring day bathed the balcony in sunshine. Sheer curtains billowed inward. A spray of white flowers blazed on the orange marble mantel. Munk ordered a pot of coffee and three bottles of San Pellegrino. He sat down and polished off a coffee, then sprang up to take a phone call at the desk, conducting his end of the conversation in monosyllables. I did not know it at the time, but as we met, Munk's company, Barrick Gold Corporation, the world's biggest gold miner, was in the midst of a secret bidding war with the Chinese goliath Minmetals Resources. One month later Barrick would stun the business world with the news that it had paid $7.6 billion to snatch an Australian copper miner from Minmetals' jaws. It was the biggest acquisition in Barrick's history, and the first time it had bought a nongold asset. Munk drove Barrick to the top in gold mining by a single-minded focus on one metal and one place. His strategy turned a $14 million gold mine into a $50 billion company. That's the happy landing. The flight itself was not so smooth.

MUNK GRADUATED IN 1952 FROM the University of Toronto with a degree in electrical engineering. Six years later he and business partner David Gilmour founded Clairtone Sound Corporation Limited. The company's mission was to produce high-quality record players, and encase them in the best design. In those days, home audio equipment came in boxlike cabinets. Clairtone introduced slick, modular components that captured the spirit of the 1960s.

The company aimed at the top end of the market. Munk hired Frank Sinatra as a Clairtone pitchman, and managed to get product placement for his futuristic Project G stereo in the iconic 1967 movie *The Graduate*. He rose to business stardom, jetting around at the age of thirty addressing business groups who saw him as a visionary. A 1967 commercial posted on YouTube shows Munk and his partner rolling across the Brooklyn Bridge into Manhattan in a gleaming 1936 Pierce-Arrow convertible. In the backseat is their latest product, a Clairtone color TV. The Clairtone company was falling apart around them when the ad was shot, but Munk was as cool and sleek as the Danish modern cabinets that enclosed his products, and that attracted customers like Hugh Hefner.

Later, overwhelmed by manufacturing problems and a move into color television not supported by demand, Clairtone collapsed. In a piece on the company and her father, Nina Munk, a *Vanity Fair* contributing editor, quoted a lawyer for the Nova Scotia government, which lost millions of dollars in a partnership with Clairtone: "Munk was too good a salesman for his own good. He could sell anything to anyone—including himself."

In the wake of the debacle, Munk left Canada for London. In 1970 he and Gilmour founded the Southern Pacific Hotel Corporation, with Munk as chairman. Within three years it was the largest chain of luxury hotels in the Pacific. The restoration of his business reputation seemed complete until, in 1973, Israel's defeat of Arab forces in the two-and-a-half-week Yom Kippur War led Arab oil producers to embargo supplies to some countries, and to raise the price of oil. The cost of travel increased and the global economy slowed. Munk's enterprise withered, and his attempts to save it led to a humiliating meeting in London with the Earl of Inchcape, chairman of the British shipping giant P&O.

"Desperate for capital, and exhausted from a 40-hour flight from Fiji," said a profile in Toronto's *Globe and Mail*, "Mr. Munk had come to beg Lord Inchcape's board to reinstate an offer for shares that his South Pacific company had earlier rejected. The imperious Lord Inchcape would not be moved. 'What offer are you talking about, Mr. Munk? There is no offer on the table. You will go bankrupt and then we will talk to the receiver and will pick up your assets and you will face the wrath of your shareholders, which you well deserve, because you were pigheaded.' With that, Mr. Munk was shown the door."

Munk managed to salvage his investment, but was burned again, this time by an oil venture that went sour when the market turned. In his telling, this was Munk's own *De Profundis*, the moment that misfortune stunned him with a reversal "so quick and so brutal and so merciless." He retreated to his home in Klosters, the Swiss resort, to lick his wounds. The year was 1982. As Munk picked through these old events in his hotel room in London, it was clear that we had reached the moment of epiphany.

Klosters, a retreat favored by Prince Charles and home to people like the banking scion Nat Rothschild, had been Munk's refuge from business cares for years. This time, though, he thought only of business. In particular he thought about the vagaries of price, and how swiftly they could demolish any plan. He thought too of his own blunder in following the wisdom of the herd. Munk was fifty-three years old—in the prime of his life. Proud and striving, he searched for an opportunity to recoup his own and his partners' lost millions. He wanted a business that could withstand a fluctuating price. "We needed to find a business before it became popular, a business that was so unfashionable no one wanted to get into it." His attention turned to a resource whose price was not rising, but falling. As Munk saw it, the world's leading supplier of this commodity, South Africa, was in a countdown to political upheaval. Best of all, the companies that mined the metal held their value even when the underlying price was falling.

"Gold, sir," Munk declared. "Gold! It carried the highest multiples. Gold shares sell at a very high value in relation to their earnings because a gold share is perceived to be not just a share but an option or a call on gold as well. If you buy Swatch watches, if you buy Nestlé, you buy the earnings. If you buy gold shares you buy it because, hey!—this company has 2 million ounces of gold and I think that gold will go up in five years!

"We had only a few million dollars left in the kitty, not more than $20 million. I said, 'Guys, let's find a gold mine.'"

In 1983 Munk and his partners formed Barrick Resources. Most start-up gold miners aim to grow by discovery. Munk had no patience for rummaging in the bush. Barrick's strategy would be to buy reserves, not find them. Growth would mean *rapid* growth. Munk built his company like Lego, snapping gold mines into place after deciding what would fit. He bought some mines for their gold and others for their people. Much nonsense has been written about Munk. In one biography he arrives for his deportation from Hungary brimming with unconcern—an anecdote that managed to be both painful and silly. Another chronicle contains a well-worn story about Munk, fresh from a dazzling performance at a shareholders meeting, stepping into an elevator, and being propositioned by a gorgeous stranger. "That's nice," says Munk in response to her offer to sleep with him, "but what would it do for my shareholders?" But Munk's story is heroic anyway. He escaped death at the hands of a killing machine. He survived humiliating failures. He built the biggest gold miner in the world with a single-mindedness and speed that demand admiration.

Barrick bought its first mine in 1984, a struggling operation in northern Ontario. To renovate the mill—an essential improvement if the mine was to be profitable—the company needed to make an immediate investment of $15 million, money that Barrick did not have. In raising the money, Munk showed the financial audacity that he could bring to the service of his ambition. He offered a new type of investment, a gold trust.

Investors in the trust did not buy shares in Barrick, but in an off-the-top percentage of the mine's future production. An "escalator" clause sweetened the deal: if the gold price rose, so did the trust's percentage off the top. Either way, the investor would make money even if the company did not, since the trust raked off its share ahead of costs. The trust gave investors a way to hedge with gold, to place a marker on the gold price without the risk of changes in the cost of operations. Hedging would become the signature tactic of Barrick's growth.

Barrick's next purchase was Camflo Mines, a 50,000-ounce-a-year Quebec producer with one enormous problem: it owed the bank $100 million. The mine's banker approved the buyout, but gave Barrick one year to pay off the entire debt. Adding to this burden were Barrick's money-losing oil investments, still carried on the books at $30 million.

First, Munk unloaded the oil properties for $32 million. He raised a further $30 million in a private placement of Barrick stock with fifty large investors, and $53 million more by floating a second gold trust. The bank recouped its $100 million and Barrick got the gold. More importantly, the company captured in that single acquisition something it had lacked and needed—a team of top gold miners.

"Until then," said Alan Hill, one of the executives who came to Barrick with the Camflo purchase, "Barrick didn't even have a staff of operating personnel. They ran their mine with consultants."

In Munk's grand scheme, Barrick would grow into a company large enough to attract the European fund managers who had always kept a portion of their assets in gold and gold mining stocks. These funds had generally placed the gold mining share of their investment in South African mines, the industry leaders. Rising violence in the apartheid country, and public sentiment in Europe and America for disinvestment from South Africa, would, Munk thought, force fund managers to look elsewhere for their traditional gold mining investment. Munk planned to give them the opportunity to invest in a large gold company in a safe environment: the United States. As soon as Barrick acquired its new staff of mine executives, Munk sent them out to find the all-important American acquisition. "We looked at every gold mine that was for sale in North America," said Hill. They found what they were looking for thirty-five miles southwest of Salt Lake City.

The Mercur mine had a fabulous past, but it had closed in 1913.

In 1981 the gold mining division of Getty Oil bought the mine and spent $100 million trying to bring it back. In 1984 Texaco bought out Getty. Since the gold mine was not part of what Texaco wanted, they engaged bankers to look for a buyer. Alan Hill went down to Utah for a look.

Hill and his team were crunching the numbers when, in January 1985, Exxon Corporation, the oil giant, appeared with an aggressive rival bid. The Barrick team left the field to Exxon and went looking for another prospect. One month later Exxon inexplicably lost interest and pulled out. Hill went back to the Mercur, and this time found what he thought was a blunder that could hand Barrick a bargain.

At first glance he'd thought that Mercur's reserves, enough to last eight years, were too low to support the cost of buying them. But as he looked again, Hill thought that Texaco might have overestimated the costs. Since ore is a function of cost as well as gold price, lower costs would mean that there were more reserves than Texaco had thought.

At this point, Hill recalled, Texaco cut him off from access to the mine. The company worried that its workforce would become demoralized to discover the mine was for sale. Instead, Texaco set up a data center in Denver. That's where Hill took his second look at the numbers and found the oil company's blunder. He did not want to tip off Texaco that it was understating its own reserves, but felt that the only way to check his calculations was to run his numbers on the oil company's own reserve-calculation software. He labeled his lower costs "interim," and asked Texaco to run the numbers. When they did, the reserves increased by 25 percent. Hill believed the true figure would be even higher. Munk went after the mine.

Texaco had set a minimum price of $40 million, a large sum in 1985 for a company as small as Barrick. Texaco, a multinational oil giant, was also reluctant to commit the security of its employees to such a tiny outfit. In a face-to-face with the executive in charge of the sale, Munk made the case that Barrick could run a mine and meet its obligations. He then baited his offer with a sweetener: if the gold price rose by a stipulated amount, he would pay Texaco an extra $9 million. The deal was done. "Peter Munk," said Alan Hill, recalling these events, "could finance anything on the planet."

For $40 million Munk got a mine that previous owners had spent $100 million improving. His experienced mine executives increased the ore flow through the mill from 3,000 tons a day to 4,000 tons, and later to 5,000 tons. In three months they reduced recovery costs per ounce of gold from $290 to $212. In less than a year Mercur's annual production rose from 34,000 ounces to 116,000 ounces, and revenues from $13 million to $42 million. Barrick had a winner, and most importantly, had it in America.

Mercur established Barrick as a midsize gold producer with a foothold in the United States. One of Munk's strategic pieces was in place: location. Now he needed the second piece: a big enough company to attract large investment funds. Barrick would have to find and buy a company with an important gold reserve. There was only one place for Barrick to locate such a reserve and remain in the United States: Nevada. As Hill put it, it was time to go hunting in elephant country. The elephant they found was a pitiful beast. They turned it into the most famous gold mine in the world.

THE GOLDSTRIKE MINE CAME INTO play in 1986 when a one-half owner needed cash. Alan Hill and his fellow executives flew down to size up the mine. They were not impressed. "Goldstrike was a ma-and-pa operation run on a shoestring," one said later, "and things were slipping between the cracks." "It was a haywire operation," Hill agreed. The on-site managers ran the mine from a dilapidated trailer. Most of the equipment was in poor condition. The cyanide tanks for recovering the gold were behind a rented office in nearby Elko. By contrast, next door to Goldstrike lay Newmont's booming mines. Compared to them, Goldstrike was in a pitiable state. Yet what if Goldstrike's operators had simply failed to exploit their asset? On every side, Newmont spilled forth gold. It was easy to imagine threadbare Goldstrike missing something. A Barrick geologist wondered whether Goldstrike's multiple shallow pits pointed to a larger reservoir below, from which the upper-level gold had come. Grasping at this possibility, in a final effort to add sparkle to the ground that it was trying to sell, Goldstrike's owners drilled the main deposit to a depth of 1,800 feet . . . and hit a monster.

The drill intersected 391 feet of high-grade rock. There *was* a deeper body. But the deposit was refractory. The gold was bound up in a sulfide lattice that would make recovery difficult and expensive. Profitability is what distinguishes ore from worthless rock. High recovery costs can strip the value from a target, and economic methods can enhance it.

GOLD IS A STUBBORN ELEMENT. It sits near one end of the periodic table, among other inert elements. Chemically it does not want to leave the state that nature has put it into, and must be dragged out by the hair, often with the help of toxic chemicals. One of these chemicals is cyanide, and a common system that employs it is known as "carbon in pulp," or CIP.

With a CIP recovery mill, the ore is trucked from the pit and piled in a storage area called the run-of-mine pad. A conveyor takes the ore and feeds it into a massive revolving drum. Inside the drum, steel balls cannon into the rock, battering it to pieces. A second ball mill smashes up whatever the first one missed. The powdered rock is mixed with water to create a slurry, and piped into the first of a series of tall leaching tanks.

In the tanks, cyanide is added to the mixture. The cyanide dissolves the tiny specks of gold from the pulverized rock. At the same time, carbon pellets are pumped into the downstream tanks and forced upstream against the flow. The carbon is charged in a way that attracts the cyanide-gold solution, which sticks to it. By the time the ore has passed through all the tanks, the carbon will have blotted up most of the gold.

Finally, the carbon pellets pass through an acid wash that strips off the gold. The cyanide goes to a tailings pond, the carbon cycles back into the process, and the "pregnant solution" of gold passes through an array of electrified metal plates wrapped in steel wool. The gold electroplates onto the steel wool. Once a week the operators pull this rig apart, sluice it out with a jet of water, and recover a fine sand of gold. They dry it in an oven overnight, then melt it and pour it into bars.

On the Carlin Trend, miners also use a cheaper method of extraction called heap leaching. The technique has added billions of dollars of value to gold miners by converting large tracts of marginal ground into ore. The practice of recovering metal by dribbling liquids onto heaps of stone is more than 500 years old, but the first use of cyanide to recover gold from a heap was at Cortez, Nevada, in 1969. Gold mining never looked back.

Cheaper than CIP, heap leaching lets operators process large tonnages of low-grade ore that otherwise would just be waste. They feed the rock through a series of crushers, then pile it into heaps hundreds of feet high. Each heap rests on an "impermeable membrane," an industrial-grade plastic sheet. A typical large heap can cover 160 acres. A cyanide solution dribbles onto the heap through a web of perforated hoses. The cyanide percolates through the crushed ore, releasing gold and carrying it in solution down to the plastic collecting sheet. The pregnant (gold-bearing) solution runs off into pools. The liquid is then pumped through carbon to retrieve the gold.

But at Goldstrike, Barrick faced a special challenge. The new ore body was composed of sulfide ore, which is not susceptible to cyanidation. Like the chocolate in an M&M, the gold is surrounded by a hard shell. Cyanide cannot penetrate the shell of sulfide rock. There are two ways to tackle the problem. You can burn the sulfide off in an autoclave, or use specialized bacteria to chew it off.

In nature such bacteria live in places like the hot springs of Yellowstone National Park, where they have been dining on sulfur for millennia. Gold miners whet the bacteria's appetite by mixing gold concentrate with nutrients. This mixture goes into tanks where the bacteria eat the sulfide casing, freeing the gold. This process may sound "green," but the bacteria create heat, which must be dissipated by expensive cooling. Autoclaves are expensive too, both to build and to run, but Barrick already knew a lot about them. The Mercur property also contained sulfide deposits. "We were very comfortable with autoclaves," said Hill. Even considering the extra expense, then, Munk could not resist the deep-ore possibilities at Goldstrike.

At first Munk's bankers balked. If Goldstrike were so great, they asked, why hadn't Newmont bought it? Surely with mines surrounding Goldstrike, Newmont understood the ground. Munk got the bank onside, but faced another threat to the deal: a fast-closing tactical window. The drill hole into the deep deposit had gone down in October. At the end of that year, 1986, a capital gains tax provision important to Barrick's calculations would expire.

In Denver, lawyers for both parties worked straight through the holidays, the deadline hanging over them. They broke only for Christmas Day. On December 26 they were back at it. At the last possible moment on December 31, "and I mean minutes before midnight," one participant recalled, the lawyers realized that Goldstrike had not properly constituted its board. The directors on hand could not execute the agreements legally. Barrick's down payment of $10 million to seal the transaction was waiting on the table. With the minute hand on the clock almost touching twelve, Goldstrike lawyers gave Barrick verbal assurance of a quick approval as soon as they could constitute the board. The check slid across the table, and Barrick got the Goldstrike mine.

Today it is hard to imagine the struggling property that changed hands that night. One day I went to look at it with Dean Heitt. We drove up a little road along Sheep Creek, looking at the hundred-year-old stone dams of the original miners. The hillside was speckled with yellow flowers. We passed an abandoned adit and came out on a hilltop in the Tuscarora range at 7,000 feet, and there under a fresh blue sky spread the panorama of the mines—the Pete and the Carlin, North Lantern, Beast, Genesis, Goldstrike, Bootstrap. Eight million ounces of gold a year was coming out of that single swath of carved-up mountainsides. Huge, flat-topped artificial hills of tailings ranged above the pits, and miles of leaching heaps. The ocher and red of the upper strata gave way to the black pay dirt below. Mine trucks loaded with gray ore beetled out of the Goldstrike pit.

Munk paid $62 million for Goldstrike and got an Elko office hung with cobwebs, mine trucks missing doors, and problems with production. When extraction costs such as the autoclaves were factored in, Barrick had paid $300 an ounce for gold that was then selling for less than $400. According to Barrick's in-house chronicler, Newmont watched with "cool scepticism" as the Barrick newcomers arrived. The only road to Goldstrike passed through Newmont land. Barrick staff called it the Ho Chi Minh Trail, after the wartime supply route used by North Vietnam. But Goldstrike's new owners were on fire. They had a million-dollar drilling budget, and conviction. "We knew we'd find a lot more," said Hill. "The previous owners had one of the best holes ever drilled. They had *not* had the money or the knowledge or the want to look at the sulfide ore. We were ready for anything. We were hungry."

Three months after moving onto the property, Barrick drilled a hole that intersected 600 feet of high-grade ore. It contained 0.36 ounce of gold per ton—a result so good that they checked it twice. The next hole was even better, and still another hole recovered 450 feet at a sensational grade of 1.089 ounces a ton. "We found a great big anomaly," said Hill, "right across [the] ore body." They drilled to the west and found the anomaly's extension. They started drilling holes closer together, to fill in their understanding of the size and richness of the target. Everywhere they put a drill, they hit ore. In one of the most intoxicating passages in contemporary exploration, Barrick's geologists were finding a million ounces of gold a month. Sixteen months after the midnight signing in Denver, Goldstrike's reserves had grown from 600,000 ounces to 15 million ounces, and with this staggering jackpot came a fresh, new, urgent challenge: finding miners, and a town to put them in.

Barrick was enlarging the deposit by the day. "We were finding the equivalent of a small gold mine every week," said Hill. They needed miners immediately. Hill knew where to find them, but where would they live? Elko was a small town with little surplus housing. Moreover, the prospective employees would bring problems of their own. "We wanted experienced people," said Hill, "and a lot of them came from mining towns that had gone bust. Those people had houses and mortgages that they'd had to walk away from. So they didn't have good credit."

Barrick created a plan to encourage banks to offer mortgages. The company bought land north of the Interstate in Elko, and subdivided it. A contractor in Salt Lake City agreed to build a standard three-bedroom house for $50,000. The cost of each lot to Barrick was $12,500, for a total cost per house of $62,500. Barrick would lend the $12,500 land cost to a miner on the understanding that if he stayed with the company three years, Barrick would forgive the sum. With secure employment and 20 percent down payments in their pockets, the miners got mortgages. In five years the value of a Barrick house climbed to $80,000, and the forgivable contribution to $20,000. The subdivision grew to 700 homes.

In less than a decade Barrick outlined 30 million ounces. Goldstrike powered the company into the ranks of the world's biggest gold producers. More than a million ounces a year came out of the mine, and still does. A year after the discovery of the Goldstrike deep zone, the gold price had climbed from $380 an ounce to $500. Yet Munk did not grow Barrick into a colossus by betting on the gold price: he bet against it. He was already offering investors protection from country risk. Now he would protect them from another menace, one that had wrecked his plans before—the whim of price.

· ● ·

IN BARRICK'S EARLY YEARS THE gold price ranged widely. It started 1985 at $300 an ounce, hit $500 an ounce in 1987, and sank to $360 in 1989. It clawed back to above $400 by the start of 1990, but three years later had subsided to $330. Nearly ruined by his earlier brushes with the oil price, Munk evolved a system that protected Barrick from the roller coaster of price, guaranteed the company's returns, and provided the steady flow of cash that Barrick used to buy more mines and expand them. He sold forward.

In forward selling, a miner contracts with a commercial bank to deliver a certain amount of gold at a specified future date. The commercial bank then borrows that amount of bullion from a central bank, sells it in the market, and deposits the money to the miner's credit in a trust account earning interest. When the miner delivers the promised gold to the commercial bank, the miner collects the money in the account, and most of the interest. The commercial bank returns the gold that it receives from the miner to the central bank, with a small amount of interest. The commercial bank makes money on its share of the interest from the trust account, less what it pays the central bank. Everybody wins. The commercial bank profits from a low-risk transaction; the central bank earns interest that it would not otherwise have earned on its reserves; the miner protects itself from a price drop because the price it gets is the price at the time it made the deal, and it was making interest all along on a product that it hadn't yet produced. Forward selling made Barrick bulletproof to a falling price.

A single input can demolish the forward seller's math—a rising gold price. Let's say a forward contract calls for delivery in five years at $800 an ounce. If the spot price is lower, fine: that's what you were hedging against. But if the spot price is $1,400, the forward seller has "lost" $600 an ounce. Such a trap snapped shut on Barrick.

By 2009 Barrick had a massive forward hedge, trailing far behind the real value of the metal. This forward position dragged on the company's share price like an anchor. As practiced at Barrick, forward selling gave the miner the option to roll the contract forward when it came due. In other words, Barrick could delay the reckoning to give the spot price time to come back down. The price did not come down. It rose. Munk's board faced the inevitable in 2009, taking advantage of a price dip to buy out the forward contracts for a bruising $5.6 billion. In doing so they acknowledged what the rising price was saying. New buyers had arrived in the market, changing it forever.

THE GOLD PRICE HAD STARTED climbing long before the banking crisis gave it orbital velocity. Two important factors drove it. One was the appearance of a class of new gold investments that made it easier to buy bullion. The other factor was the rapid opening of a market. It was a market that wanted and bought everything—BMWs, French wine, iPods, hamburgers, diamonds, an aircraft carrier, and inevitably, gold. It was the dream market. It was China.

7

LINGLONG

To become the biggest gold producer, China had to subdue
the natural animosity of communist ideology to a substance
so quintessentially the stuff of private wealth.

IN A TEN-YEAR EXPLORATION BINGE, CHINA GREW
from an insignificant gold producer into the world's biggest, replacing South Africa in 2007. They were buying gold too. In 2012
Thomson Reuters GFMS, the world's leading precious metals consultants, predicted that Chinese private and government buying
would amount to almost a thousand tons of gold that year, a rate
of consumption that would make it a bigger gold buyer than the
traditional leader—India. The story of Chinese gold tells the story
of the Chinese state—its transformation from an inward-looking,
xenophobic, technologically backward country into a money mill.

They have 10,000 gold mines, or maybe 60,000. I was told both.
No one seems to know the number. Chinese gold is as much a frenzy
as an industry. Some mines are no more than a single family with

gold pans and a stretch of river; others exploit large deposits from Burma (now Myanmar) to the Mongolian border. All together, they produce more than 300 tons of gold a year. I wanted to see these mines, especially the fabled Linglong mine, and flew to Yantai on the Yellow Sea.

The night I arrived, Tropical Storm Meari came ashore and battered the Shandong Peninsula. Rain poured through faulty seals around the windows. The lights went out. I stood at my thirty-eighth-floor window and peered out. I could see taillights crawling on the coastal road, that was all.

In the morning everything sparkled and glittered in the fresh, cool air. Packs of black clouds charged across a bright sky, dragging trails of rain. Happy at the prospect of visiting the ancient goldfield, I devoured breakfast in my room, put the tray aside, and turned to my prize—a big yellow book on the gold mines of Shandong. My attention settled on a picture of an imperial minister setting out for Linglong in 1007. He rides a white horse, his gold cape trailing behind him. The emperor watches him depart. A retinue of miners, loaded with spades and baskets and carrying sacks of tools, trudges through a forest. At ten o'clock my cell phone rang and I went down to meet my guides.

They arrived in a tiny purple car, dodging a pile of storm detritus in front of the hotel and pulling to a halt across two parking slots. A large and cheerful woman named Pang Min beamed at me from the wheel. The backseat was stuffed with children's toys. Wedged among them was Feng Tao, a solemn gold geologist who had flown in from Beijing to be my interpreter. Feng had come to me through an introduction, and had already made me understand, with some firmness, that without such introduction all of China would be shut against me, and certainly that part of it that contained the splendid and chaotic Shandong gold industry. Pang Min herself was the fruit of this same process. And, so Feng made clear, her sizable daily fee was a trifle to be swallowed without complaint.

"Fine," I said as I folded myself into the front seat, "but we need a bigger car."

"Yes," said Feng. "She has agreed to change it."

We shot away from the hotel with every sign of purpose. There followed half an hour of spirited navigation among the abandoned streets that surrounded the hotel. Buoyant with gold revenues, and anticipating growth, the local government had filled an enormous tract with eight-lane thoroughfares, a riverside park, two stadiums, and a convention center. The parking lots were empty. Nothing was going on inside the stadiums. Streets ran for miles through empty scrub. Happily there was no traffic, so that when Pang Min coasted to a puzzled stop in the middle of this concrete prairie, we were not in anybody's way. At last we found our way out, got a bigger car, and set off westward through the sodden countryside to see the celebrated Linglong mine.

Linglong gold mine lies in the jumbled granite of the Shandong Peninsula, China's storied goldfield. Probably a hundred other gold mines thread the nearby hills with narrow, dangerous tunnels. Many are centuries old, and small. China produces more than 300 tons of gold a year yet does not have a single large mine. The story of Chinese gold production is a parable of China's transformation. To become the biggest gold producer, it had to subdue the natural animosity of communist ideology to a substance so quintessentially the stuff of private wealth. Linglong's story is a tale of changing hopes and fortunes driven by the evolution of the Chinese state. As the state changed, so did its posture to gold mining. From suspicion of gold as a symbol of private wealth threatening the communist ideal,

China has concluded, as its growing bullion holdings show, that gold might be a useful hedge in a shaky-dollar world.

All morning the rain came and went, sweeping in from the sea in thick black packages of weather that drummed on the car roof and promptly blew away. There were torn-up trees along the road. We left the main route and climbed through a slot in the hills. Pine trees cloaked the hillside. The storm abated. The tops of the pink granite slopes vanished into a mist. We drove past a line of pale yellow houses and arrived at the famous mine. A cantonment of white buildings stood against the mountain. We parked the car and climbed out and Pang Min pointed up the hillside to the place where the main tunnel entered the rock. The scene looked less like an industrial site than like a Chinese landscape brushed onto a scroll.

IN ANCIENT TIMES, THE LEGEND says, a river flowed into the sky from the peach orchard of the Grand Old Lady of the West to a peak in the Linglong hills. The river fed streams that plunged back down the mountain and collected in a basin. One day a hunter discovered the pool teeming with gold fish. They swarmed through caves and leapt from the surface spewing streams of golden liquid. News of this pool reached the imperial court. A court official who arrived in Shandong to report on the fish was so overcome by the sight that he flung himself into the pool and disappeared. A second emissary arrived to find the pool dry. Terrified of returning to the Forbidden City with such news, he drafted 10,000 peasants to dig in the mountain for the fish. They found the Linglong gold.

The Chinese were mining gold by 1300 BC. They had systematic ways to prospect for it. In the seventh century an official wrote, "If the upper soil contains cinnabar, the lower will contain gold." They had also discovered that plants could point to precious minerals. "If in the mountain grow spring shallots, there will be silver under the ground; if leek in the mountain, gold." They may have been right. Plants do take up minerals from the soil. In 1980 researchers at the University of London found plants that could take up gold and accumulate it in their tissue. A 1998 article in *Nature* described an experiment where plants absorbed gold from polluted ore. In 2004 a New Zealand researcher proposed a system of "phyto-remediation" in which plants would harvest metals from toxic gold mine waste. The payoff—two pounds of gold per acre. And a paper published at the 19th World Congress of Soil Science, in 2010, said that even today some prospectors "use plant species as bioindicators of the presence of gold in soil." The Chinese noticed such relationships in ancient times.

They had novel ways of refining gold, such as feeding it to ducks. First, they removed as many impurities as they could from panned gold, mixed what was left with chaff, and stuffed it into the ducks. Later they collected the droppings, panned out the gold a second time, mixed the refined material into another batch of chaff, and down the hatch. They did this three times, put the recovered gold aside, then fed the ducks a final batch of chaff—pure chaff, this time—to scrape out any gold bits left behind.

At the beginning of the first millennium China's rulers owned gold reserves of 200 tons, about the same as the Roman Empire's gold stock at the time. Then the mining collapsed to a pitiful trickle. When the Mongols came to power in the thirteenth century they tried to turn the gold mines around. They suppressed bandits and put down petty tyrants who were robbing the mines, and warned their own generals not to steal gold. To revive the industry, one

Mongol emperor ordered 4,000 households in Shandong to pan for gold.

In the seventeenth century Linglong came into the hands of the famous eunuch Wei Zhongxian. Originally a criminal and a hopeless gambler, Wei had traded in his testicles for a post in the imperial household. By tradition, many domestic positions in the court could only be filled by eunuchs. Wei's creditors would not be allowed to pursue him into the Forbidden City. Once in the palace, he befriended the nurse who cared for the crown prince. When the boy became emperor at the age of fifteen, Wei began his climb to power.

Wei murdered opponents, built shrines and statues to himself, and adopted the title Nine Thousand Years, almost equal in dignity to the emperor, known by custom as Ten Thousand Years. The emperor gave Wei the Linglong mine in 1621. He ran it for six years, and the country's gold production increased to 40,000 ounces a year. When the emperor died, Wei's enemies strangled and disemboweled him. Seventeen years later the dynasty collapsed, and so did Linglong.

The new Manchu rulers hated gold mining. They believed that gold veins were dragons' veins, and digging them up would disrupt feng shui, the system of laws that governed spatial harmony. More importantly the Manchu feared unrest among the people they had conquered, and precious metals could finance an insurrection. They decreed an end to gold mining. They executed those they caught disobeying the law. Without new production, the imperial treasury wasted away.

The last gold bars were sucked from the Chinese treasury in 1842, in reparation payments to the foreign occupying powers—Russia, Germany, France, Britain, and Japan. Desperate for cash, the ruling Dowager Empress repealed the gold mining prohibition. Annual production rebounded. By 1888 China was producing 700,000 ounces a year, ranking it fifth in world production.

Plainly, miners had never stopped mining. They just went elsewhere for the market. A British colonial official in Burma, writing when the Chinese ban on gold mining was still in effect, noted that "by far the larger portion of all the gold used [in Burma] is brought from China. It is imported in the form of thin leaves of gold, made up into little packets, each packet weighing about one viss." A viss equaled 3.6 pounds, a lot of gold in any century. And that was just one packet. Gold had leaked out of China along the Himalayan smuggling routes. When the gold mining prohibition ended, the gold simply headed to the nearest market, the domestic one.

Such shifts in the official attitude to gold have been repeated in modern times. Communist China's posture toward private gold has changed from approval to disapproval and back again, in time to the convulsions of the day. Yet Linglong should have had an honored place in the Chinese government's mythology. They captured it from a hated enemy, the Japanese, who had seized it at the start of World War II.

In constructing my account of Linglong, and for much else on Chinese gold, I relied on a remarkable book, *The Gold Mining History of Zhaoyuan with a Review of the Gold Industry in the P.R.* [People's Republic of] *China*, a meticulous chronicle in English and Chinese. Written by Gong Runtan, a thirty-three-year veteran of the Shandong gold industry, and Zhu Fengsan, the cosmopolitan mastermind who helped propel China to the forefront of world gold production, the book supplies, to my knowledge, the fullest account of Chinese gold available in English. From the use of gold in pills to promote immortality to a review of the Ninth Five-Year Plan, the volume has it all. It also has a gripping style, as in the story of Linglong's wartime fate, when Japan captured the mine.

"The accumulated snow on Mount Linglong had not yet melted. The biting wind sent withered branches and leaves flying across the sky, producing shrill howling.

"Early in the morning a group of confused farmers, bringing along both the old and the young, swarmed out of their home[s] and ran off into Mount Linglong. The crack of guns came near, bullets swept above the crowd, thick smoke columns rose in the village in the distance and curled up on the [dirt] road."

Seven hundred Japanese soldiers, supported by bombers, marched into the hills and occupied the mine. They fortified the mountain with blockhouses and rings of electrified fence. They built machine gun nests and patrolled the mine with what the Chinese called "wolf dogs." They brutalized civilians: the bodies of massacred Chinese, eviscerated by bayonet, hung from telephone poles around the mine.

"Linglong under the iron heel of the Japanese invaders," Gong wrote, "became a hell on earth with demons and monsters . . . and evil spirits running amok, and an exceedingly brutal concentration camp. The miners led the tragic life of the vanquished."

In six and a half years Japan took 290,000 ounces from the Linglong mine. The pages that describe it shake with Gong's indignation. I wanted to meet him, and since he was Pang Min's father-in-law, she knew where to look. At two o'clock on a sweltering afternoon we drove into the courtyard of a gray apartment complex in the heart of Zhaoyuan, and trooped upstairs.

Gong was then seventy-five. He showed me to a seat in front of a low table set with a platter of sliced watermelon. The apartment was small, but the ceiling was high and a breeze stirred through the cool, terrazzoed rooms. A mass of plastic flowers bloomed madly on a wall. Gong sat straight, with his hands on his knees. His skin was yellow, and he looked hot. He wore his iron-gray hair in a crew cut. He wore frosted lenses that concealed his eyes.

Only two years old when the Japanese invaded, Gong had researched his account of Linglong by talking to men who had worked at the mine under the Japanese. One of the men he interviewed had infiltrated Linglong for the communists.

"He helped steal gold from the Japanese," Gong told me through Feng Tao. "There were many of them. That was their job. They were sent by the party. The party had a lot of farmers in its membership, so the men looked like farmers, and the Japanese hired them. In fact they were guerrillas. They had a hidden smelter at Linglong where the gold was poured into molds and cooled in water. They shaped it into bars that could be sewn into clothes and smuggled out."

It obviously gave Gong pleasure to think of Linglong in a patriotic light. A moral relish infused his writing. A collaborator named Jiang Qixu, for example, "was steeped in iniquity, with hands stained with the people's blood all over." Sadly, the literature of mining does not abound in such accounts.

The occupation of the Linglong mine ended in style in a storm of gunfire, with the cornered Japanese commander defiantly flinging his sword at a Chinese officer, who shot him dead. The collaborator Jiang was dragged from hiding and executed. "The crisp shot echoed among the Linglong Mountains."

The retaking of the Linglong mine marked the end of the world war, but four years of fratricidal combat remained as the Chinese fought each other to a finish in the civil war. Although Linglong's production fell, the mine contributed 7,000 ounces a year to Mao's treasury. When the communists won the civil war in 1949, Linglong might have been written into a triumphant history—a mine from imperial times turned to the service of the people! Instead, gold fell under a pall. In the wars of ideology that followed, even to be associated with the metal was to risk contamination.

GOLD EMBODIED THE IDEA OF personal wealth. This presented communists with a dilemma: was gold a blessing for the desolated postwar state, or a threat to socialist purity? In 1957 this question seemed to have been answered when the premier, Zhou Enlai, backed a directive aimed at "organizing broad masses of the people to produce gold." The broad masses were probably already at it, but now it would be legal, part of a plan to revive the mines and stimulate the production of a valuable resource. No sooner did this initiative appear, however, than it was swept away in the purges that convulsed China: the Anti-Rightist Campaign, the Four Clean-Ups, and the scourge sometimes called the war of the young against the old—the Cultural Revolution.

Approved by Mao Zedong to root out what he saw as capitalist sympathies in the leadership, the movement sent brigades of young Red Guards into the country to hunt out "deviationists." In a decade-long rampage, gangs of zealots brutalized the educated, ruined the legal system, destroyed families, raped, murdered, maimed, and burned. In this poisonous milieu, gold was a target too. The Chinese characters for gold became a synonym for the hated bourgeoisie. Geologists stopped looking for gold. The China National Gold Group Corporation, the country's main producer and itself a creature of the state, was labeled a "bourgeois trust," and dismantled. Officially, the Chinese gold industry ceased to exist.

∙ ● ∙

ONE OF THE MEN WHO later transformed China's gold industry into the powerhouse it is today was at first purged and persecuted himself. When I met him, Zhu Fengsan was eighty-two. Erect and trim, he had thick hair combed straight back, and his characteristic expression was a wry, appraising smile. When I visited he was recovering from an accident. He'd crashed his new electric-powered bicycle. In the collision he shattered his patella. He pulled up his pant leg so I could see. We studied the yellow bruise and eight-inch scar in silence for a moment. "They think it was my fault," he said, "because I am in my eighties and should not be trying new electric bicycles."

The youngest son of a Shanghai banker, Zhu was born in the city of Hangzhou. Marco Polo called it "beyond dispute the finest and noblest city in the world," and the Chinese agree. "Above is heaven," goes the saying, "below is Hangzhou."

Zhu went to a high school attached to Saint John's University, a prestigious Church of England institution in Shanghai. After high school he attended the university for a year, studying architecture. He hoped to finish his studies in the United States, but in the Chinese civil war he could not get a passport. After the communist victory, university admissions were restricted, and the courses limited to subjects considered useful to the state. Zhu picked geology, and graduated from the top-rated Tsinghua University in 1952. His fiancée and classmate, Zhou Mingbao, also a geologist, graduated at the same time. While Zhu was wondering what to do with himself, Emily, as he always called her, took a job as a teacher.

"At that time," Zhu said, tapping my knee for emphasis, "*no* one wanted to be a teacher! Emily was the only one. She was the *only* one who applied." He sighed and shook his head. "I loved her, so of course I was obliged to follow."

They moved to Changchun, a long train journey northeast of Beijing, and took up posts at the Geological College. Six years later, in 1958, the Red Guards sniffed him out.

As he told me his story, Zhu never criticized his country, and always called the communist victory of 1949 "the liberation." Yet however blandly he described events he must have suffered anguish. He had been dean of his faculty for four years when the Red Guards tore him from his family and exiled him to the countryside to be "reeducated" by the peasants. He worked on roads and labored in the mines. He returned to teaching in 1962, but in 1969 he was denounced again. This time his wife and children went with him, and for three years toiled on a farm.

When the state allowed him to return to teaching in 1972, it still considered him politically unreliable. His school would not let him teach about strategic minerals, such as uranium. Gold, on the other hand, was ideologically inert. Lenin had defined its place in socialist thinking in 1921 when he said, "When we are victorious, we will make public toilets out of gold." Even a doubtful character like Zhu could be allowed to teach about such a lowly metal. "And right away," he said, unfolding his elegant hands as if revealing a discovery, "I found it interesting."

Mao died in September 1976. One month later a coup toppled the Gang of Four, the powerful executors of Mao's policy. The reformer Deng Xiaoping took power. Gold returned to favor as a mineral with obvious attractions to a country eager for foreign exchange. In 1978 the gold price averaged $200. The next year it hit $520. Spurred by the prospect of potential gains, the Chinese military established a gold prospecting unit. The Gold Army, as this new corps was called, grew into a force of 20,000 gold explorers. Zhu became adviser to the top Gold Army leaders. In three years he jumped from the backseat of Chinese geology to the front. His rise mirrored China's swift U-turn. Deng abolished class-based discrimination. The banker's son was not only rehabilitated, but in the midst of a booming, reinvigorated industry. Gold officials commanded large forces and broad powers. What they lacked was knowledge.

The transformation of China from midsize producer to the top of the gold mining world is a story Zhu helped write. By touring foreign mines and organizing a symposium for Western gold geologists, he opened the door to the technology that has helped China unlock its reserves. In its strategy for exploiting this newly fashionable resource, China demonstrated an extractive genius far ahead of its abilities in the field: a knack for mining Westerners.

8

THE BANDIT CIRCUS

It's the richest gold deposit in China. We found it, we
defined it, but they just were not going to let us keep it.
—Colin McAleenan

IN 1993, TO ENCOURAGE EXPLORATION AND EX-
pand its gold reserves, China loosened regulations and opened the
door to foreign miners. Even Chinese state-owned companies could
now raise exploration money by selling shares to private investors.
One large foreign gold miner to take advantage of the opportunity
was Placer Dome Inc., an international miner based in Vancouver.
Placer Dome had been doing business in China since the 1960s,
supplying copper and molybdenum to China National Nuclear Corporation. In 1997, through its Australian subsidiary, the company
hired two Chinese geologists, formed Placer Dome (China) Ltd.,
and started looking for gold.

In 1997 China was the world's fifth-ranked gold producer.
Largely unexplored by Western standards, China looked like a rosy

prospect. Not only had the country loosened restrictions, but the all-important Five-Year Plan, the government's regular economic manifesto, made the restructuring of the mining industry a key objective. Placer Dome had a five-year plan of its own: train Chinese staff to higher standards; demonstrate that "large mineral systems" existed in China; and get the clear title they would need to develop them. They did train up their staff, they found a target, and in five years had their business license. When it was time to lock up title, here's a partial list of what they learned:

- three different levels of government can issue title

- not all title information was on the same register

- access to the registers was limited

- access to state geological data was limited

- topographic data was a state secret

- reserve calculations were not based on international market standards

- owners' share of revenue would be "modest"

In the end Placer Dome got nothing but the doubtful pleasure of transferring expertise to China. It was a pleasure the Chinese were anxious to supply to others, as Colin McAleenan discovered.

McAleenan, an Irishman, had studied geology at Trinity College Dublin. When he graduated, he worked in far-flung parts of Canada, until he settled in Vancouver. He was an experienced field geologist and exploration executive when in 1994 he made his first trip to China to scout for targets. In the next few years he examined a gold prospect in Szechuan and traveled to Inner Mongolia to look at properties.

In 1999 McAleenan visited the glittering port city of Dalian for China's first international mining conference. Backing him were a pair of American investors: Robert Kiyosaki, the creator of the best-selling *Rich Dad Poor Dad* series of motivational books; and Frank Crerie, an Arizona-based financier who had made fortunes in uranium and oil.

With the Dalian mining conference, the Chinese government hoped to attract foreign mining capital to the country. McAleenan represented money looking for ground to drill. "It was the first year they'd held the conference," he said. "They wanted to create the hottest mining event in Asia, but at first it was poorly attended. They gave technical talks on China's mineral endowment and the country's open-door policy, but there were only two or three booths where provincial government entities were showing off their projects." One of these properties caught McAleenan's eye. It belonged to the Bureau of Mineral Resources of Liaoning, the province Dalian was in. "They had paper on a number of projects, and one was intriguing. I brought it away with me, and we thought about it, and then I came back for a look."

McAleenan flew into the provincial capital of Shenyang and took the expressway down the spine of the Liaodong Peninsula. The road was one of the first of China's now ubiquitous multilane highways. It had been built by Bo Xilai, the mayor of Dalian and one of China's fastest-rising politicians. Bo's career blew up later in a scandal fed by corruption charges, Communist Party infighting, and his wife's alleged involvement in the murder of a British businessman. But in 1999 he was in the forefront of China's speed-of-light emergence into the modern world, and the expressway symbolized the sense of opportunity and commercial vitality that China's leaders wanted to project.

The terrain of the peninsula was rough but not mountainous. A sparsely covered sawtooth range of hills extended into the Bohai Gulf. McAleenan reached the exploration target, at a place called Maoling, or Cat Hill. The provincial geology bureau had drilled some holes on a quartz-vein target. "They were using pretty old equipment," McAleenan said. "They had these massive Russian drills that could drill a vertical hole hundreds of meters deep, but that's it. If the vein dipped steeply [descended at an angle] they couldn't intersect it. It was antiquated."

McAleenan took rock samples from outcrops, including material from the veins and between the veins. He took the samples back to North America for assay, and waited to see what he would find. When the results came back, he was amazed. "Not only were the narrow veins rich, but the material between the veins had gold too."

McAleenan thought that the best way to tackle the deposit would be to mine the whole hill in an open pit bulk operation, instead of trying to follow the veins. Such a method would result in a lower grade overall, but since capital costs and operating costs were low in China, and he expected the gold price to rise, he believed such an operation would be profitable.

On behalf of the American financiers and Mundoro Mining Inc., a company set up to exploit Chinese targets, McAleenan negotiated a deal with the corporate arm of the Liaoning province minerals department. In July 2001 the parties formed a joint venture to exploit Maoling. Under the agreement, Mundoro would earn a 79 percent share of profits in return for funding and directing all the exploration and carrying the project to production.

For two years Mundoro worked flat out. They drilled sixty-two holes, pulled 20,000 meters of drill core from the hill, and trenched for 13,000 meters. They analyzed the soil chemistry, translated earlier

Chinese documents about the site, and sank some pits. In the end they had enough evidence to report to the market a deposit containing 1.1 million ounces of high-probability "indicated resource," and more than 4 million ounces "inferred," meaning ground that they expected to raise to the "indicated" level with more drilling. In November 2003 they took the company public and raised $13 million.

An eager wind was blowing through the Chinese gold scene. The open-door policy to foreigners had sent junior exploration companies fanning out into the country looking for property to drill. One of the most hyped of these was the Boka prospect, a target in Yunnan. Villagers had discovered gold in the hills; the regional gold brigade had drilled it; a Vancouver junior called Southwestern Resources Corp. had swept in and grabbed the license. "Junior" means a small mining company focused on exploration. Southwestern drilled sixty-seven holes and estimated a resource of 150 tons. At the grades they claimed, the deposit would hold 5 million ounces. Promoters started calling Boka the highest-grade gold deposit in the country, and talked about a windfall of $40 million for the local economy. Southwestern's market capitalization grew to more than three quarters of a billion dollars. "The problem," said an official, after several years went by, "is that at the moment, nobody knows when the first piece of gold will be found."

Alas, somebody did know. Southwestern's Canadian CEO knew. He knew that the first piece of gold wasn't coming up anytime soon, because he'd been falsifying the drill results. Southwestern's share price later collapsed and in 2013 the executive went to jail, but in the heady Chinese gold rush scene of 2003, with the whole country flung open like some fantastic oriental Yukon, the news about the Boka strike was pumping helium into everyone's balloon. "It probably helped our financing," McAleenan said.

What helped even more was that all of McAleenan's holes were hitting gold. "We couldn't miss. We had long intersections [where the drill was passing through gold-bearing rock]." Mundoro's home page showed a panorama of the site, and if still there it's worth a look. The ore body stretches from a little loop of road cut into the top of Cat Hill on the left, all the way across the hills to the right and out of frame—a strike of half a mile. You can see the exploration tracks hacked through the vegetation. The gold price was rising. Mundoro's share price was rising. After two years of intensive exploration Mundoro went back to the market for more cash. They raised another $25 million. The shareholders were confident. McAleenan was confident. The American financiers behind Mundoro must have been confident too. Then the business license came up for renewal. It didn't get renewed.

The date came and went. At first there seemed no cause for worry. "We sent our people to ask why we didn't have the license," McAleenan said, "and got answers like 'It's nothing—just the way things work in China,' 'Things move at their own pace,' 'They're always late,' so we didn't worry. But when months passed, we started to get concerned. We could get no action from the Liaoning province officials, so we went to Beijing."

There were two licenses Mundoro needed. The first was the local business license, and the second was the exploration permit. To continue work at all, that permit was essential. Issued in Beijing, the permit called for a regular accounting of money spent on exploration. "We had exceeded the minimum requirements," McAleenan said. "We'd drilled 40,000 meters. We'd done environmental and geotech studies. We'd advanced the project on a number of fronts simultaneously. We had developed two scenarios: one for a 30,000-ton-a-day mill and one for 50,000 tons. We were calling it the largest undeveloped gold deposit in China, if not in Asia. Our share price at the initial offering had been $1.25 and now it was more than $4.00."

In Beijing, Mundoro went to the land resources ministry and asked if their permit would be a problem. No, said the officials, it would not . . . as long as Mundoro got the business license. Without the license, the ministry said, Mundoro did not exist legally.

You can see where this is going: McAleenan's dawning realization that they were going to get screwed; the appeals to diplomats in Beijing; trade attachés getting cold-shouldered at the ministry. Up in the Liaodong Peninsula provincial bureaucrats were sitting on the business license, and if they were sitting on it, someone had told them to sit. Mundoro's title to the property, like Mundoro's joint venture, just "did not exist."

McAleenan left the company at the end of 2006, after a year of futile attempts to preserve Mundoro's entitlement. Five years later the company managed to recover $13.5 million, the amount raised by the initial stock offering. Instead of 79 percent of profits, they were reduced to 5 percent.

"If it had been a lead-zinc deposit," McAleenan said, "I don't think we'd have had a problem. But the political leadership decided they wanted to keep the gold for themselves. It was a national asset. It was the biggest gold deposit in China and they were not going to let a foreign-controlled joint venture have it. I think that 9-million-ounce deposit is going to be more like 18 million. It's the richest gold deposit in China. We found it, we defined it, but they just were not going to let us keep it."

At Maoling, China won two prizes: the gold mine, and the skills that had developed it. They'd had a front-row seat while their partners drilled out the deposit. I think it's fair to say they decided they could take it from there, using a skill set that they did possess—the operation of an impenetrable bureaucracy. In the Chinese gold rush of the day, most Western companies went home unsatisfied. Toxic Bob did not.

WHEN CHINA MADE THE 1993 decision to encourage mine development, it allowed any Chinese company, no matter its core business, to explore for any metal, and in 1995 China National Nuclear Corporation dug up some ore on a lonely ridge in Inner Mongolia called Chang Shan Hao. China Nuclear had known there was gold in the area because artisanal miners had been scratching out a living. The company excavated a pit, made three test heaps of ore, dribbled cyanide onto them for thirty-two days, and recovered 65 percent of the estimated gold content. Then they went looking for a Western partner.

The first two companies to examine the site decided against it. Then it caught the eye of Robert Friedland, the billionaire American minerals magnate known as Toxic Bob for a 1991 cyanide-and-heavy-metals leak at his company's mine in Summitville, Colorado. The cyanide techniques at Summitville had delivered profitable gold yields from low-grade ore. Chang Shan Hao presented a similar opportunity, and in May 2002, using a Toronto-listed junior called Jinshan Gold Mines, Friedland bought the property.

"Frankly, the property was not properly explored when Robert came in," said X. D. Jiang, the mine's general manager, a diminutive geologist, and the real genius of Chang Shan Hao. "There were about a hundred artisanal miners still working on the site. It was a mess."

Friedland took it over anyway and hired Jiang in 2002. Jiang built a pair of fifty-ton test heaps. He experimented with crushing the ore more finely than China Nuclear had crushed it, produced a feasibility study, and in February 2007 completed a recovery plant that could process 20,000 tons of ore a day. They piled the ore into heaps, leached out the gold with cyanide, fed the pregnant solution into tanks containing carbon in solution, and stripped out the gold. In July they poured their first 500-ounce bar. Other mines were pouring too, and one year later, in 2008, China overtook South Africa as the world's biggest gold producer.

The year that China moved into first place, Beijing whisked away the old minerals policy and replaced it with another. Gold became a restricted commodity, one that the Chinese preferred to mine themselves. Friedland was looking for money anyway, to cover the "cash burn" at his Oyu Tolgoi gold-and-copper strike across the border in Mongolia. In May 2008 he sold his stake in Chang Shan Hao to the state-owned China National Gold Group for $200 million. Friedland called his relationship with China National Gold "strategic." It certainly was for the Chinese. They kept a connection to the man who was developing the richest mineral play in the world, Oyu Tolgoi. Eventually Friedland's joint venture partner in Mongolia, the giant Rio Tinto PLC, muscled him from the board, but that would not have mattered to the Chinese. They owned almost 10 percent of Rio too.

On the face of it, Chang Shan Hao was a doubtful prize—a scruffy, low-grade pit. In most parts of the world, it would struggle to be profitable. But if ore is a product of math, in Inner Mongolia the math was right. The province had the world's biggest coal deposits. With the coal, they made cheap power. With cheap power, they made cheap steel. Cheap power and cheap steel and cheap labor put the sparkle into Chang Shan Hao.

• • •

INNER MONGOLIA'S FOUNTAIN OF COAL has showered the province with millionaires. In a 2012 table of China's million millionaires (people with net assets of more than 10 million renminbi, or $1.6 million), Australia's *Business Review Weekly* put the Inner Mongolia total at 13,500. Many of them seemed to be staying at my hotel in the industrial city of Baotou. The ground-floor bar blazed with gold Rolexes. Women in stilletos tottered through the marble lobby with shopping bags from the high-end-label stores next door. Outside, a red Maserati glowed among the ranks of jet-black SUVs. A corps of stony-faced drivers with mirrored glasses waited for their bosses. A dusty white Toyota Land Cruiser came roaring up the drive and pulled to a stop beside me. I climbed in, and we sped away to Chang Shan Hao.

The road north of Baotou climbs into orange mountains. From the heights you can see the green haze of fields along the Yellow River. First we drove on a four-lane highway that wound through foothills. Every few miles the driver slowed to a crawl, crept past a parked police car, then shot back to his normal speed. In a series of sprints we made our way through the mountains and onto the plain until, clear of police, the driver accelerated to breakneck speed.

For half an hour we sailed along the motorway. I could see the green glow of agriculture far off to the left. Suddenly, without warning, the driver wrenched the wheel and shot off the freeway onto an unmarked dirt track that plunged into a dry creek bed. We bucked along through gravel and fine sand, throwing up red plumes that broke across the hood. It must have been a well-known shortcut. A line of large blue trucks with loads that swelled out at the top like loaves of bread came toward us. We tossed cinnamon-colored clouds of silt onto each other's windshields as we bounced by.

We reached pavement, and struck off along a highway lined with poplars. Sad-looking, crumbling hamlets appeared and receded in the distance. A Chinese village clustered at a crossroads around some engine shops and a truck café. We came to a railroad construction site. A ridge of gravel ran like a ruler across the plain from one horizon to the other, with a notch left for the highway.

The humps of Mongol burial mounds appeared on the steppe, and sometimes a more elaborate painted-concrete tomb with a shallow dome and a forecourt surrounded by a low, white wall. I thought of the Mongols riding out of this plain in the thirteenth century— the most devastating cavalry the world had ever seen. Now native Mongolians fear that mining, and the migration of ethnic Chinese, will annihilate what remains of their nomadic life. A month before I visited in June of 2011, Mongol protests brought riot police into the streets of the provincial capital, Hohhot, after a Chinese truck driver ran down and killed a Mongol herder who was trying to block a coal truck convoy from crossing grasslands. China executed the truck driver, but the convoys have not stopped.

Twice in two hours I saw distant herders driving animals beneath the dark sky. Black clouds had been building for an hour, and then a squall dashed out and drummed on the roof. At a highway tollgate a young Mongol couple, their bicycle loaded with luggage, sheltered from the rain. Beyond, the highway dipped down to a river swollen by the rain. Chunks of mud sailed across the road in a frothy slick. The ditches dissolved into rushing streams. We crept along with the wipers flailing, and came to a pirate mine.

The site was abandoned in the rain. A front-end loader and a truck were parked beside a hovel. Smoke oozed from a stovepipe and dribbled down the wall. Sluices and a sifting screen stood by the river, with ramps for the trucks to bring ore. The miners had diverted the river, dredged the bed, and had been washing gravels down the sluices to collect grains of gold. They had messed up the river for miles, and would certainly have bribed local officials. In China I often saw miners foraging in rivers. Once in Shandong I watched a highway crew with a backhoe scooping out a stream to patch the road, and while they were at it, running gravel down a sluice.

Near Chang Shan Hao illegal miners had been dynamiting hills, crushing ore, and piling the gold-rich soils into leaching heaps. Paper sacks that looked like cement bags were stacked beside the ore. The bags contained sodium cyanide briquets. The miners watered the cyanide from local streams and let the solution seep through the heap to liberate the gold. Cheap plastic sheets beneath the heaps caught the gold-bearing cyanide precipitate. The pirates collected the concentrate and trucked it to small refineries in Baotou that process the material and cast it into bars.

We came to an expanse of grassland bordered by distant hills. A ruined fort stood near the road. Crumbling towers flanked the gate. The ramparts ran about a thousand feet in the long dimension. Evidence of China's ancient contest with the Mongols is printed on the surface of the province. Old walls appear and disappear in the desert. On my Chinese map they were all called "Wall of Genghis Khan."

The fort was another residue of that old contest. A cluster of Mongol houses and some goat pens stood beside it. I watched a Mongol girl lash a stubborn goat with a knotted cord. Her skirt flew out and her arm flashed and the goat tried uselessly to dodge the blow. A mile down the road we drove through the gold mine gate.

A ROW OF SAPLINGS STRUGGLED in the wind. On a hillside, a painter balanced on a ladder diligently brushed a design of boatmen onto the dome of a pagoda. China Gold had a policy of improving the appearance of its mines. On the slope above the pagoda stood a yurt, the traditional leather Mongol tent, ostensibly a nod to the country's past. But they were short of housing at the mine, and the yurt was full of Chinese workers.

A path led up the hill to an observation point. From Chang Shan Hao the plain stretched west into Central Asia. I thought of all the travelers who had set off through those deserts. In 1935 Peter Fleming (brother of Ian Fleming) made a seven-month trek from China to Kashmir. In *News from Tartary* he described his encounters with isolated Chinese governors, Tibetan pilgrims, a Mongol prince, and the small parties of gold washers, itinerant placer miners scouring the streams of Inner Asia. Such bands had picked for centuries at the surface expressions of huge deposits now coming to light.

Chang Shan Hao had nineteen drill rigs going on a narrow strike three miles long and 200 yards wide. The miners had already followed the ore down to 500 feet, and the deposit was "open to depth," meaning that the deepest drill hole had not found the end of the ore. The drill plan called for a further 175,000 feet of exploration holes. The mine expected to double its reserves from 168 million tons of ore to more than 300 million tons, or about 5 million ounces of gold, and a mill expansion is now forecast to increase output to 260,000 ounces a year by 2015.

We drove up to the pit on a road that circled the mine, and at a platform that surveyed the scene we got out to take a look. Red and yellow ore trucks swarmed the pit. Lines of vehicles loaded with ash-black ore crawled up the ramp while empty ones went down. Even at a glance, something seemed not right. After a moment of watching the activity, I realized what it was. Compared to a Western mine, pit traffic was inordinately dense. The open pits I was familiar with were austere, even solemn in appearance, not like the teeming crater at Chang Shan Hao.

AT A MINE LIKE GOLDSTRIKE, where a truck hauls 370 tons, pit managers keep traffic at a stately pace. They never crowd the mine. At Chang Shan Hao the pit floor writhed with trucks. This activity was a function of truck size. Compared to such behemoths as those at Western mines, the Chinese trucks carried only fifty tons. To keep the plant in ore, they jammed more trucks into the mine. When we drove off to tour the mill we met a line of brand-new yellow mine trucks trundling to the pit, and I commented to Jiang that they looked like mine trucks anywhere. "Oh yes," he said, "they look the same. But they break in a month."

The same was true of Chinese pumps. Jiang had American pumps at Chang Shan Hao that had run continuously for two years. The Chinese pumps were much less durable. Using such equipment was not a flaw in the business plan, but part of it. In a cheap-steel environment the cost-effective choice was to buy the local products, and when they broke, buy more.

Labor was also plentiful and cheap. At Chang Shan Hao a workforce of a thousand produced, when I was there, 110,000 ounces of gold a year. Typically a miner earned $650 a month. The mining contractor was China Railway, the biggest company in China. Notoriously corrupt, it consumed one percent of China's entire GDP. It had more than 2 million employees, its own police force, court system, schools, hospitals, and telecommunications. It has a culture of military discipline from its origins as a department of the army. At the mine, China Railway workers slept eight to a four-bunk room, swapping bunks at the shift change. They kept their quarters in immaculate order. They worked long hours in hard conditions far from home. As a mine executive observed when I remarked on this, "there are always others waiting at the gate to take the place of anyone who does not like the work."

X. D. Jiang, the master of Chang Shan Hao, lived in relative splendor. He had a suite at the end of the hall in the managers' wing of the motel-like living quarters. His bedroom was about four times the size of any of the identical rooms that filled the rest of the wing. The legate of a powerful state company, China Gold, and the mine's top manager, Jiang received marks of respect that no Western mine boss would expect. Mine staff sprang to their feet when he entered a room.

Jiang seemed never to stop working. I went to see him one night at his office on the second floor of the administration building. At the end of a dark corridor, light spilled into the hall from the corner office. Three brown-leather sofas formed a U that faced Jiang's opulent lacquered desk. A golden dragon about three feet tall snarled from its plinth. In a map on the wall, China occupied the center of the world. The diminutive Jiang sat swallowed in his huge black leather chair, wreathed in cigarette smoke—as isolated by his position as once the governor of the nearby fort had been, separated from those around him by authority, education, and background.

The eldest child of a Chinese literature professor, Jiang was born in 1958 in Weihai, a city in Shandong on the Yellow Sea. He was nineteen in 1977 when Deng Xiaoping restored the system of university entrance exams. Jiang won a university admission, and in the practice of the time, was told where and what to study. He took the four-year course at the Changchun Geological Institute, and graduated in 1982.

For the next twenty years Jiang put his drills into targets from China to the Arctic. He worked at mines in Ontario and East Africa. In 2002 he returned to Vancouver, where he'd settled his family, arriving just as Toxic Bob was buying Chang Shan Hao. Friedland hired Jiang, and Jiang set off to see if he could make a gold mine out of the heaps of low-grade rubble piled on the lonely hill.

Jiang arrived at a ragged site that had a few test heaps and a pit. Pirate miners rooted through the property. The resource that Jiang wanted to assess was one that demanded meticulousness and patience, and certainly resolve. The business culture of the province was rapacious, even for China. Coal millionaires sneered at the concept of a low-grade deposit, particularly one that needed so much careful assessment. They preferred a commodity that could be gouged out of the ground in bulk and converted into cash almost on the spot. Gold took too much time.

Illegal miners scouted the geology of the deposit and infiltrated the recovery operations to see how Jiang treated the ore. They took this knowledge into the hills nearby and started the noxious digs and leach heaps that were dribbling cyanide into the watercourse.

Local government officials tried to enrich themselves from the mine by extortion. They trenched across the main road and demanded fees to fix it. "In China," said Jiang, when I asked him what he did about it, "the first thing you must find out is who your enemy's boss is. Once you know that, the problem is not so hard, because then you know who *your* boss must call." An icy phone call from Beijing straightened out the highway issue.

Many stories have documented Chinese corruption and the stupefying riches of the "princelings," as descendants of China's revolutionary leaders are known. But when it comes to mining gold, the whole of China is a bandit circus.

• • •

CHINA IS THE WORLD'S BIGGEST gold producer not because it has large mines, but because it teems with small ones. According to Greg Hall, an Australian geologist and gold mining executive, and a director of the listed company through which China Gold runs Chang Shan Hao, China has some 11,000 registered gold mines. In 2011 the total annual production at the country's ten biggest mines was less than fifty tons. The remaining 270 tons, then, came from the other 10,950 mines. These thousands of mines are therefore mostly tiny. What's more, the number of such micro-mines might be even greater than the figure Hall reports. X. D. Jiang told me he saw a government document in 2004, when authorities were trying to eradicate illegal mining, that put the number of unlicensed gold miners then at work in China at 60,000. Such a number would beggar belief were it not that the government itself had created the boom.

In the 1980s, before China opened exploration to Western companies, the government tried to stimulate production by lifting state control of small mines. Many small miners were already placer mining in the country's goldfields, and the new policy sparked a gold rush. Gold seekers ransacked prospective ground in every corner of China. They poured into such desolate immensities as Qinghai province in the Tibetan uplands, and the Xinjiang Uygur autonomous region, the largest administrative division of China: 641,000 square miles of deserts and mountains and gold-bearing streams. Not all of the new miners took to these distant wastes. Some descended on goldfields closer to home and started stealing ore from existing mines. And there were other problems.

Contemporary gold extraction uses complicated processes that consume energy, material, and demand specialized knowledge. But people have been refining gold for centuries without the help of engineers or substantial capital. Probably the most common method is mercury amalgamation.

Mercury dissolves most metals, forming solutions called amalgams. One way to create a mercury-gold amalgam is to place mercury in the riffles of a sluice and wash the crushed ore down across the riffles in a stream of water. Riffles are grooves that run across the direction of flow, like the ridges in corduroy. As the ore flows over these, gold particles "amalgamate" with the mercury: they stick to it. Refiners can pick out the gold and reuse the mercury. The problem is that mercury poisons those who handle it, and ruins the environment. And mercury amalgamation is not the only toxic method that the Chinese used.

On a prospecting trip in mountainous Hunan province, on the Burma border, X. D. Jiang saw artisanal gold miners using sulfuric acid—an insanely hazardous technique. "We came across little valleys filled with the white smoke of acid fumes," he said. The hill miners used cyanide too. Men on motorcycles carried drums of cyanide up mountain tracks. The miners built leach pits downstream from the digs and poured in cyanide. It leaked, poisoning people and causing widespread cattle death.

In 1988 the Chinese government reversed its liberalizing policy. Gold became a "special mineral" under government protection. Yet the genie would not return to the bottle. For many Chinese, the idea had taken root that gold alone could free them from penury. In 1989 the *Christian Science Monitor* profiled the sad adventure of a farmer named Ma, who set out on a March day to seek his fortune in the goldfields of Qinghai province. "Ma hired a dozen men, spent his life savings on shovels, tents, food, and a truck, and left his ripening barley to the hoeing of his wife." The men drove out of their village, crossed the Sun and Moon Mountains, and struck out westward across the high plain of Qinghai.

Ma and his crew were following a stream of outlaw prospectors and gold diggers who were ignoring the prohibition on private mining. The lure of gold was too great to resist, and the get-rich-quick promise of the "Crazy Yellow Rush," with its stories of instant millionaires, drew them to the remote quarters of the province. Local officials also defied the law, helping the miners and pocketing substantial bribes and "fees." One group in the central Qinghai city of Golmud reportedly made $270,000 selling bogus mining licenses.

Ma and his team plunged doggedly across the steppe. They traveled nearly a thousand miles on the rough roads that led to the gold country. Rain fell and the snow blew in and the roadway turned to mud. They ran into other gold seekers mired in the sucking mess. More than 8,000 men and 1,000 trucks were tangled in hopeless disarray. They had reached an altitude of 16,500 feet, within sight of the famed "Gold Platform" at Hoh Xil Lake. And then the food ran out.

News of the calamity reached the authorities, and the army tried to make a food drop. But the pilots could not find the stranded miners, concealed from the air by thick cloud. More than seventy men died in the freezing pass. Finally the army reached them with helicopters, bringing food. Ma and the others were ordered to abandon their doomed convoy and leave the area on foot.

The disaster in Qinghai did not eliminate small-scale gold mining. The fever had taken root. Moreover, small mining was an immemorial tradition. In acknowledgment of this the government found a way to tolerate small workings, under an unwritten but well understood convention known as "localistic protection"—a Maoist concept that gave weight to the folkways of the people. The people knew an opportunity when they saw one.

In Henan province, 178 small-mining teams used the cover of localistic protection to seize ground adjacent to state-owned mines. Some brazenly entered mines and collected ore at the minehead. In an incident at the Wenyu mine, the "carrying-ore-on-the-back people" had gathered so much ore they had to commandeer trucks to haul it away. The mine manager, desperate to conceal his helplessness from an impending state inspection, arranged with the looters for a three-day suspension in the plundering while he showed the government around.

Sometimes freelance miners organized into village cooperatives, a system known as *min cai*, or village mining. Even large state companies used *min cai* to recover hard-to-get-at ore, in one case suspending miners on ropes and lowering them 120 feet down shafts. In Shandong province *min cai* provided important income. But in 2001, in another shift of central government sentiment, the state abolished the system, declaring 295 mines and 138 refining operations illegal in Shandong alone. The government ordered more than 7,000 migrant workers from other parts of China to leave the province.

Officially, such micro-mining no longer exists, and gold recovery has passed to the industrial apparatus. This fiction does not withstand even the most cursory investigation. As we drove through the goldfields of Shandong, I badgered Feng Tao to prevail on Pang Min to show us one of the small operations. She finally threw up her hands, turned off the main highway, and took us down a little road through fields and orchards until we came to the quartz hills. The road ended at a midsize gold mine, and I thought that Feng Tao's efforts had foundered on his difficulties with Pang Min's Shandong accent. But then we saw it, in a field not a hundred yards from the gate of the modern mine. Steam poured up through the rusty, antiquated superstructure of a hoist that rose above a corrugated fence erected to conceal it. The volume and density of the vapor plume testified to the depth of the mine. Rock temperature increases with depth, and torrents of water were needed to cool the rock face. The water changed to steam and billowed back along the mining gallery and up the shaft. About a mile away another column of white vapor billowed up through the rickety derrick that marked a second shaft.

Official hostility to small operations comes partly from the role they play in providing cover for the theft of ore from the surrounding mines. A truckload of ore in the possession of a gold miner, however small, is not evidence of theft, because such miners often transport their ore to refineries rather than refine it themselves. With the connivance of insiders, thieves steal from one mine and sell to another. They may even sell ore back to the mine they stole it from. If the price is right, a mine may prefer to buy back pilfered ore rather than lose it entirely. In Shandong, where there are so many buyers, selling such stolen ore would not be hard.

In Zhaoyuan city alone there are six large refiners, but Pang Min knew that I was interested in the smaller, freelance side of the action. After nosing around in the hills we returned to the city and stopped at one of the industrial refineries. Pang Min chatted to the guards until a young man emerged and climbed into the backseat beside Feng Tao. We pulled away and plunged into the busy streets in another of Pang Min's exuberant displays of navigation. Finally we sped across a bridge and into a suburb. Pang Min and our new companion kept up an animated dialogue. "What's up?" I asked Feng. "I don't know," he replied, "I can't understand a word." We turned off a wide street and drove up a muddy lane until we came to a crumbling ten-foot-high brick wall. On the other side, towering above the wall, was a grayish yellow mountain of tailings.

We turned in through rusty gates and stopped by a decrepit lodge. Some men came out in shorts and T-shirts and plastic sandals, and our new guide spoke to them. Feng and I got out and took a look around. It was a concentrating mill. Ore from small mines in the hills, or stolen ore from larger mines, came down by truck to be crushed with millstones and refined into concentrate in flotation ponds. In flotation, the refiner adds an oily dark brown liquid with a pungent smell to a slurry of crushed ore. The chemical attracts gold-bearing sulfides into a froth on the surface of the pond, where operators harvest it as concentrate and sell it to industrial smelters, where they burn off the dross in large furnaces.

CHINA BECAME THE BIGGEST GOLD producer in the world because the country, taken as a whole, is an effective machine for sucking gold out of the ground. Part of the machine is the industrial gold mining structure. The other part is the rapacious mass that the government let loose in 1978 when it encouraged them to look for gold, and has never been able to recall. When farmer Ma and 8,000 others trudged away from Qinghai's Gold Platform in defeat in 1989, there were thousands more to take their place.

In 1995 "gold lords" and their bands of peasant miners were still defying the government, and each other. "I've heard the bullets whizzing past my ears and seen people beside me dying," a woman in her forties told a reporter. "Even if the government wanted to control us, we are stronger than they are right now, and our guns are normally much better than the ones carried by [police]."

Gold lords fought for the best ground. "The first thing we would do when we struck gold," one said, "was set up a pillbox on the mountaintop and train more of our workers how to use guns; then we would use our blankets to make a flag, which meant: this mountain belongs to us."

Rival gangs fought pitched battles with pistols, automatic weapons, and even homemade cannon. They chewed up the landscape with their mining, sending soil from the digs into rivers, threatening the habitat of such rare animals as snow leopards and white-lipped deer. In 2012, nomadic Tibetan shepherds in Qinghai fought with Chinese gold miners to keep them from digging a mountain that the Tibetans held sacred. Nothing, it seems, can stop the sack of China for its gold.

ONE DAY I DROVE OUT of Yantai through the orchards to look at a pit mine on a feature called the Jiao Lai basin. Overloaded trucks with squealing axles gasped up the ramp from the mining level. The pit was a shambles, the slopes scarred by gravel slides. The fence sagged along the ground. Illegal miners, the manager admitted, looted the property and the entire Jiao Lai basin more or less nonstop. In the cost conditions that prevailed, the low-grade mine and the plundering around it made a good contraption for extracting gold, as you could say for all of China. But it begs the question: with such intense mining, how much gold can be left?

In 2010 the World Gold Council warned that "China's future prospects for gold supply look to be limited." The WGC is a research and marketing organization run by the gold mining industry. Using data from the United States Geological Survey, the WGC said that "China may exhaust existing gold mines in six years or less if Chinese demand continues to grow strongly."

For anyone in the gold business, other than Chinese miners, this was great news. China running out of gold? Imagine what would happen to the price.

The China-running-out-of-gold scenario raises the specter of "peak gold," the scary bedtime story that gold producers like to tell. Peak gold describes the theoretical point of maximum world extraction, after which it's downhill all the way as production fades until we've dug up every ounce and there's no gold left to mine. I don't know how much gold China has in the ground and neither does the WGC. Neither do the Chinese. But if they show signs of running out, the gold price will reach escape velocity. What it will escape is human reason.

9

THE SPIDER

Every three months the London market was trading twice
as much gold as had been mined in all of history.

GOLD IS A MAD BAZAAR. THE VOLUME OF BULLION
coming to market today is more than sixteen hundred times what it
was in the sixteenth century, when Spanish treasure fleets were jos-
tling up the river to Seville. Not only are people buying all this fresh
gold, they are buying it and selling it and buying it again.

In 2011, the total value of mined gold was about $143 billion,
yet the value traded was much larger. In one three-month period
alone, 11 billion ounces of gold worth $15 trillion changed hands
in London. The true size of this market only came to light that year.
Gold movements are secretive. Most of the trade is in the hands of a
cabal of banks in London. Those very banks led the way in exposing
the size of the market. Their reason for uncovering the volume of the
bullion trade was to provide it with a whole new class of customer—

banks. As part of a campaign to promote the standing of gold as a "liquidity buffer," a type of capital asset that European commercial banks were required to hold, the bullion market wanted to show European banking regulators how big the gold market was. First they had to find out themselves.

The London Bullion Market Association polled its members about the size of daily trading. Most bullion trades in London, the world's oldest and most important market, are on a principal-to-principal basis. The transactions are private to the parties involved, and leave no public footprint. When the size of this trading emerged, it was astonishing. In the first three months of 2011, the value of gold traded was $15,200,000,000,000. In volume terms, the figure represented 125 times what the world produced in a year, or twice as much gold as had been mined in all of history.

These trading volumes helped cause rapid movements in the gold price. New investment vehicles make it easier to trade bullion. No longer does the speculator have to buy gold bars from a dealer. He can pick up a phone and buy shares in a bullion fund. The biggest of these funds appeared in 2004, only seven years before the London bullion market discovered its own size. The people who created the fund opened a crack in a dike that had been holding back an ocean of demand, and the money poured in. In a few years they had more gold in their vault than China's central bank. Bullion analysts began to watch the gold flows in and out of the fund with as much attention as they watched such crucial market indicators as jewelry demand. In the gold world they call the fund the Spider.

The Spider is an exchange-traded fund, or ETF, an investment composed of a basket of assets that trades with the ease of trading stock. In this case the asset is a single one: gold bullion. The nickname "Spider" comes from the fund's full name, SPDR Gold Shares. (The first-ever ETF was called Standard & Poor's Depositary Receipts, or

SPDR.) The nickname fits. The Spider's customers could react with arachnid speed to the slightest tremor of investor sentiment. By removing the headaches of picking up physical gold and finding and renting vault space, the fund advanced gold ownership more than any measure since the creation of gold money in the seventh century BC. It democratized bullion. Anyone with a few hundred dollars could take a gold position. More importantly, and what gave the Spider fangs, so could anyone with a hundred million.

The Spider was a creature of the World Gold Council. The gold miners who ran it wanted a better way to sell gold, as they were producing so much more of it. From 1975 to 1985, world gold production ran between 40 million and 50 million ounces a year. New technologies, such as heap leaching, brought more reserves into the production stream. By 1995 world production had leapt almost 50 percent to about 72 million ounces, and ten years after that, to 80 million ounces. In thirty years the amount of gold coming to market had doubled.

As the gold supply ballooned, the WGC's New York office looked for ways to expand the customer base beyond the jewelry market. They took a simple approach: they canvassed American investors who did not own gold, and asked them why they didn't. "The very strong message we got back," said the executive who ran the survey, "was that gold investment was cumbersome, costly, and complicated." The investors wanted something that would be easy to trade, did not have to be vault-stored by the buyer, yet represented exposure to the spot price of gold. In short, they wanted to be able to bet on gold the way they could bet on stock. In 2000, the WGC set about inventing a product that wrestled the complexities of the bullion market into a simple form, yet one that would satisfy the strictures of the Securities and Exchange Commission.

"They hadn't a clue how the gold market worked," a WGC manager said of the commission. "I would go down to Washington with a lawyer and a gold trader and someone who had dealt in ETFs, about five of us all together, and we'd be ushered into a cavernous conference room and we'd sit around an enormous table. We'd start to talk about how the gold market worked, and you'd notice that people were filtering into the room, and at the end I'd be addressing forty or fifty people. There was a tremendous interest in how this exotic product worked."

To license the ETF for sale, regulators had to be convinced that every share would represent real gold bullion in a vault. In 2004, the SEC licensed the Spider. For the gold believer, that warm feeling was now a phone call away. The Spider went on sale in November 2004. In four days it took in $1 billion. Eight years later the Spider owned about 1,300 tons of gold worth almost $70 billion. There was more gold in the Spider's stash than in the central bank of China.

The Spider ETF made bullion ownership available to ordinary investors, who apparently wanted to own gold once the difficulties of doing so were removed. But ease of ownership added more to the gold market than additional owners. It added instability.

THE SPIDER DESTABILIZED THE GOLD price in two ways. One way was by making it so easy for people to buy and sell large amounts of bullion, and the other was a structural consequence of how the fund worked.

In analyzing the Spider's effect, it would be useful to know who its investors are, so as to understand the relative sophistication behind the trading. But there are no data to show who owns the Spider. Some of the owners' positions show up in 13F filings—mandated quarterly declarations by American institutional investors that report their trading. For instance, 13Fs revealed that billionaire George Soros sold $650 million worth of Spider shares in the last quarter of 2010, while another billionaire, John Paulson, whose managed funds had $4.5 billion in the Spider, stayed put. But 13Fs document only those funds under institutional management, products sold to the public. If Paulson or Soros had a billion dollars of his own in the Spider it would not show up in a 13F. Jason Toussaint, a WGC executive, told me that 13Fs were his only window into ownership. He could not even say how many owners had the shares, and therefore could not estimate the average size of holdings. This means that there is no way to gauge investor sophistication. We can assume that some of them are not as sophisticated as Soros or Paulson.

The Spider (and the other gold ETFs that sprang up in the wake of its success) contribute to volatility by supplying a larger public with the means to make quick bullion trades. Presumably the moves of the more sophisticated practitioners influence some of this trading. In this view, the less sophisticated investors pile on, hurling themselves into the action in an imitative frenzy. ETF trading did not drive price action as a primary mover, said bullion analyst Sterling Smith, but made its movement "more intense."

The structural effect of ETFs on price instability arises from the requirement that ETFs "rebalance" themselves at the end of the trading day, buying or selling their underlying assets to match up with the day's buying or selling of the fund. As rebalancing occurs, a flood of trading orders hits the market at the same time. The market reacts as it does to any large order, moving up if the order is to buy, down if to sell. Since ETF shareholders may have acted in response to a market movement in the first place, their actions intensify that movement.

In a DealBook column in the *New York Times*, Steven Davidoff, a law professor with expertise in financial regulation, saw in the general buzz about gold the elements of a pricing bubble in which "speculation is aided by the financial revolution. Previously gold could be bought by retail investors only through dealers and street shops. Now anyone can go on the Internet, click and buy gold in the market through exchange traded funds." In a bubble, wrote Davidoff, television and the Internet would play a large role in spreading the hype. And that is what they were doing. In such an atmosphere, he said, "the marketing of gold to the masses is an ominous sign."

You can order gold as easily as groceries. Some sellers will deliver to your door. One such company is Goldline International, Inc., of Santa Monica, California. Goldline's most famous pitchman is Glenn Beck, a popular right-wing talk show host. In 2011 he had a show on the Fox television network. Beck stirred up his pitch with forecasts of ruinous inflation. He advised his viewers to put their faith in "God, gold, and guns." In a broadcast I listened to online, Beck predicted the collapse of the European Union, and European war.

The fortunes of Goldline soared as the gold price rose. In late August 2011 gold hit $1,917.90 an ounce. The WGC reported that demand for a single category of gold investment—bars—had doubled in the previous two years to 850 metric tons a year. Television business commentators predicted a gold price of $2,500 an ounce, and gold bug websites vibrated with reports that John Paulson, with billions of dollars worth of bullion in his funds, was prophesying $4,000 gold. Leaving aside the holdings of gold coin speculators and private bullion owners, and the uncountable lumps belonging to those who'd melted down their jewelry in home smelters, the amount of investor gold in ETFs alone stood at 2,250 metric tons. Then in September 2011 the gold price reached the end of a ten-year bull run, and fell off a cliff.

In a month gold tumbled $300, landing with a thud at $1,600 an ounce in October 2011. Some investors saw their gains evaporate. On November 3 the Santa Monica, California, city attorney filed a nineteen-count criminal fraud complaint against Goldline, alleging the sale of overpriced gold coins to people who thought they were buying bullion. Goldline agreed to refund as much as $4.5 million to buyers, according to reports, and to pay $800,000 into a fund to settle future claims. ABC News reported that Goldline also had to stop telling customers that the government wanted to confiscate its gold.

As the gold price fell, hedge funds started "unwinding" gold positions to raise cash to cover the increased collateral required by the funds' bankers when other assets, such as stocks, started falling too. This overall decline of asset values challenged the rule that the gold price rises when equities fall. This time the whole ship was sinking. Analysts call it a "risk-off," in which investors jettison any risk at all, and head for cash. "Cash" meant only one thing: the U.S. dollar.

You could take the view that selling gold to raise collateral for wilting stock positions demonstrated gold's utility as a hedge. If, for example, gold had withstood the risk-off better than some other asset, the prudent investor would choose to take a loss on gold in order to escape the greater loss of selling that more depleted asset, an asset that, if held, might recover later. Moreover, the ready market for gold, even as its price declined, showcased one of the metal's undoubted qualities—liquidity: the ease with which it was turned into money.

As it happened, this quality was crucial to a coup the World Gold Council was attempting to pull off. They were asking European banking regulators to include gold in a special class of top-flight assets. If gold were included in this class, the banks would need to own it. It seemed a strange time to be making the pitch, with the gold price tanking and the markets feeling faint. But they have sturdy hearts at the WGC, and just the metal, they would say, to stabilize the teetering banks of the European Union.

AFTER THE BANKING CRISIS THAT followed the collapse of Lehman Brothers in 2008, the Basel Committee on Banking Supervision, an obscure but powerful caucus of G20 finance ministers, issued new capital requirements for banks. These requirements included bigger liquidity buffers. Such buffers were to be highly liquid assets that could provide cash in an emergency, when normal sources of cash, such as other banks, dry up. Top-quality government debt—German bonds or U.S. Treasury bills, for example—would qualify as highly liquid assets. In the submission of the World Gold Council, so would gold, and on a bright day that heralded a spell of Indian summer for London, where I was living then, I went down to the old financial quarter, known as the City, to hear why.

It was just after noon, and the steps of St. Paul's Cathedral were crowded with people sitting in the sunshine eating sandwiches and tilting their faces to the sun. I strolled through the gate into Paternoster Square. A well-dressed crowd was flowing out of the London Stock Exchange and into the restaurants around the square. If anguish about sagging markets was troubling the financial heart of London, they had decided to forget it while they ate.

I made my way into Newgate Street. Lawyers wearing wigs smoked on the sidewalk outside the Old Bailey, London's main criminal court. Down the street from the dark old court stood the entrance to a modern, glass-walled office building, and in a third-floor conference room I sat down with Natalie Dempster.

Dempster, a brisk Scottish economist, worked for the Royal

Bank of Scotland and JPMorgan Chase before joining the WGC, where she heads government affairs. When we met, she had just returned from Brussels, where she'd been pitching European banking regulators on gold's inclusion as a liquidity buffer. She slapped a wad of lobbying material onto the table and began to enumerate the reasons European regulators should like gold, picking her way through the criteria with practiced skill, and explaining how gold met them:

> *Criterion:* The asset cannot be an instrument issued by the institution holding it as a buffer.
>
> *Gold:* No one "issues" gold. Its value is not tied to anybody's credit. Therefore it has no credit risk.
>
> *Criterion:* The asset's price should be easy to calculate from public information.
>
> *Gold:* Gold is probably the best known asset price in the world.
>
> *Criterion:* The asset must be tradable in markets with a large number of participants, a high trading volume, and market breadth and depth.
>
> *Gold:* The London gold market alone was trading $240 billion a day.

It's worth repeating that the true size of London bullion trading had surprised even those who'd uncovered it, the London Bullion Market Association. It was the first time in sixteen years they'd polled their members about trading. They'd conducted the exercise in their own interest. If regulators accepted gold as a liquidity buffer, sales to banks would increase. If liquidity was the measure of a buffer, the LBMA survey revealed, gold met it. The liquidity was a function of the intensity of trading. If volume was the standard, Dempster said, it was easier to sell gold than government bonds. There were always buyers—the definition of liquidity.

Because the need for bigger liquidity buffers stemmed from the problems banks had faced when money got tight, Dempster reviewed gold's performance during the crisis. She contended that the bullion market stayed robust while other markets faltered. A graph showed interest rates spiking as banks stopped lending to each other. The bullion market stayed liquid. Many other markets "assumed to be deep and liquid proved to be the exact opposite," she wrote in a supporting document, "and assets could only be sold at a large discount. This was even true of some AAA-rated assets: credit ratings proved to be no guide to liquidity."

In Dempster's view, gold also benefited from a pricing floor. "What happens to gold is that the structural demand of the jewelry market exists under it," she said. "What happened during Lehman's failure is that suddenly, as gold fell, industrial and jewelry demand cut in and put a floor under it. So it's not about gold never falling, but that when you are liquidating it the price is never going to fall out of bed. This is one of the most important points. People from the jewelry and technology sectors have a completely different perspective. When gold goes down in price, they *want* it."

Central bank gold buying is another box that gold advocates like to tick, and Dempster ticked it. The Chinese central bank had been expanding gold reserves by about a hundred tons a year, and Mexico, Russia, and South Korea had all bought large amounts. Now European central banks seemed to be changing their position on bullion. They had been sellers for more than a decade, shedding about 400 tons a year. That trend had stopped. They had become net buyers.

In pitching European regulators, Dempster and the WGC were

trying to change decisions already made by the Basel Committee of G20 finance ministers. The committee had already considered including gold as a liquidity buffer, but had rejected the idea, mainly, a spokesman for the British Treasury told me in an email, "due to the volatility of its price." Moreover, he added, the Basel Committee's recommendations were global. European regulators were supposed to be transposing them into law, not "watering them down" in any way.

In the end they did not water them down. It was the wrong time to be recommending gold as any kind of pillar. Volatility was roiling the bullion market. September 2011 was a chaotic month for the gold price, with sharp swings up and down. Whatever winds were blowing blew the gold price. Gold did not behave like any kind of safe haven. It behaved like what it had become—just another derivative. It was a construct that could be attacked, and someone attacked it.

THE GOLD PRICE WAS VULNERABLE because it was easy to manipulate. Gold traders understand that liquidity is a sea with different depths. When the London market opens, a lot of gold is available to trade. When it's closed, much less. If you want to sell a lot of gold in an orderly way that will not disrupt the price, you sell when London is open. But what if you *do* want to move the price? Then it makes sense to pick a time when the trading is thin.

Large miners watch the markets like hawks, because they sell their gold there. They understand to a fine degree the consequences of trading in a thin market. Barrick Gold's research department once determined, for example, that 100,000 ounces offered during London market hours would move the gold price down $4. That same

100,000 ounces sold when London was closed, would depress it $10 to $15. When half a million ounces dropped like a bomb on September 7, 2011, into the thin market before the London opening, then, it was no accident.

The attack took place during a period of volatility. In the chaotic month of September 2011, the gold price dashed up and down the chart seeking a consensus that it could not find. Into this jittery milieu, on September 7, during lunch-hour trading in Shanghai, hours before the great gold blotter of the London market opened, someone dumped 500,000 ounces. The gold price dropped like a shot crow.

Suspicion turned to Libya, where the regime of Muammar Gaddafi was disintegrating in a bloody rebellion. A former Libyan central bank governor warned that the dictator had possession of the country's gold reserves. As it turned out, Gaddafi's gold sale had already come and gone. He had cashed in some thirty tons of bullion the previous April, selling it for a reported $1 billion to dealers who had gone to Libya to transact the deal. Gold traders in Tripoli's Old City, near the Libyan central bank, said that the regime had started the selling with a trickle of 22-carat coins and then increased to twenty-six-pound bars as its crumbling army demanded pay. But that was five months before the Shanghai sale.

To see what might have caused a bullion seller to make a move so harmful to the price, I spoke to Jim Mavor, an eighteen-year veteran of Barrick's gold trading operations. Mavor has since become vice president of finance at Detour Gold Corporation, but when we spoke he was still Barrick's treasurer.

"Well," he said, "somebody with 500,000 ounces of gold worth [at the time] $900 million, may be assumed to be somebody who knows what's what. So it's fair to wonder why he performed a trade guaranteed to drive the gold price down. It fell $50. In trying to understand what happened, it's useful to construct a scenario in which the person causing the event to happen made money out of it."

Such a scenario, in Mavor's view, could work this way: Let's say the seller of the gold also owned a certain kind of option called a put. The owner of a put option has the right to sell the underlying asset—in this case gold—at a stipulated price. Let's make the price $1,800 an ounce. If the put owner also happens to own bullion, in addition to owning the right to sell it, then the $1,800 put has placed a floor beneath his possible losses. He knows he can always exercise the put at $1,800 no matter how far the price drops.

But in the scenario we're envisioning, gold has not dropped, but risen. It has reached $1,825. The owner of the gold has made money on his bullion, but in the process his puts have become useless. There is no value in a right to sell gold for $1,800 when the price is $1,825. The puts, then, represent a loss, because the owner paid money to buy them. The only way his puts can regain value is if the gold price falls. If the put owner decides to drive the price down to revive the value of his options, the best time to do it is in the low-liquidity doldrums before the London opening. The only math required to make his attack on the price appealing is if our hypothetical gold player can make more money from selling the puts into an alarmed market than he will lose from the depressed value of his bullion. We must suppose he can do the math. The likely trader, in Mavor's estimation, was a hedge fund, where aggressive tactics and large sums come together.

If the September 7 bullion dumpers meant to spook the market, they picked the right time. It was already spooked. Europe was in danger of unraveling in what *The Economist* called "the greatest crisis to befall the European project in its history"—the dilapidated euro. In the United States, the contest between parties over what to do about the national debt had become a war of religion. In one view the economy would founder without government action, including a tax hike on the rich; in the other, such interference amounted to the destruction of the republic. Meanwhile, American stocks lost $1.1 trillion in a four-day rout. The U.S. government was about to run out of money. Bitter partisans haggled in the Congress. Into this reaper of despair went the gold price. In a single week it lost $200.

Confidence in gold was crumbling in its biggest market, futures. With futures, investors bet on where the price will be at a stipulated forward date. To open the contract, the buyer must deposit with the commodity exchange a certain percentage of the contract's value. This collateral gives the exchange something to seize if the market turns against the player and he is tempted to abandon his contract. In the face of increased volatility in the gold price at the end of September, CME Group, formerly the Chicago Mercantile Exchange, the world's biggest commodities market, raised its collateral demand for gold by 21 percent. It was Chicago's third margin hike for gold in a matter of weeks. The aggregate increase, according to a client note from one large bank, was *90 percent*, forcing investors who couldn't post that much collateral to liquidate their gold positions. The liquidations were depressing the price.

Also preying on the gold price were suspicions about ETFs. The gold market plays in a skeptical arena. Gold bugs are habitually suspicious: mostly of paper money and governments, but anyone will do. You don't have to search far to uncover doubts about whether gold ETFs actually possess the physical gold they are supposed to. Perhaps it is only natural for a product such as the Spider to attract suspicion, given its spectacular success. Before the market turned against gold, the Spider was the biggest ETF of any kind.

In September 2011 the fund owned 1,232 metric tons, or about 40 million ounces. All of it was stored in the London gold vault of the fund's custodian, HSBC Bank. Since the ETF was ultimately a creature of gold miners, you could say it had simplified the flow of gold from one underground space to another. Or could you? Was the gold at the end of the stream as real as the gold at the beginning? That question troubled some people's minds.

Online at spdrgoldshares.com, a visitor can navigate to a photo said to show the bars held in the Spider's account. The Spider says that its shares represent real bullion in a real vault. Such gold is said to be "allocated." In an allocated account, the asset is not merely produced when the account holder asks for it, like money in a bank account, but stays in its repository all the time, unused by anyone else. It is as if you deposited cash in a bank, and the bank put that actual cash into a separate box and kept it just for you, separate from the cash of every other depositor. No matter what happened to the other deposits, *your* deposit would be intact. When the Spider fund says that its gold is allocated, it means that the custodian keeps it separate from any other gold. It is always and only the Spider's gold.

The Spider's site also gives a list of every bar in the ETF's account, with each bar's unique number, its gross weight, and its "fine" weight—the actual weight of gold that it contains. The bars are said to be refined to London Good Delivery standards—a trade definition that specifies a purity of at least 995 parts per 1,000. But rumors persist, about gold ETFs in general, that they do not hold enough gold to redeem all shares. In this view, a surge in redemption demands would collapse the funds, as they would not have the bullion to meet the calls. These rumors, said the client note referred to above, were also helping to push down the gold price.

"The biggest gold and silver funds are now on the defensive," wrote a commentator in a piece appended to the note, "as they may soon face mass investor exits on the back of heavy discounts to the precious metal spot prices [i.e., a falling market] and doubts about the levels of physical gold they actually hold."

September 2011 closed with words like "bruising," "brutal," "turbulent"—and those were all in the same piece. In the worst quarter in three years, the markets had turned into a kind of *Friday the 13th*: you got chopped just for being there. If gold was a safe-haven asset, the sound we should have heard was the vault door slamming shut as investors sealed themselves in with the bullion. Instead, the mood of alarm did not drive frightened money into gold, but into that paper shelter scorned by gold bugs—the U.S. dollar.

GOLD LOST ITS STRUCTURAL BALLAST when it lost its formal relationship to money. Now it tosses on the same sea of events as other assets. It doesn't occupy a special asset class ordained by history: all you can say is that it did.

In 150 years the world supply has grown from 10,000 metric tons to 170,000, from a modest cube six feet a side to a dazzling block that would cover the infield at Yankee Stadium. Every three years we pack as much fresh gold onto the block as our ancestors mined in 6,000 years. We trade it like men possessed. Every quarter an amount of gold equal to twice the amount ever mined flies back and forth in London in a storm of trades. And that's just the metal. Trading alongside it is an even thicker blizzard of derivatives. Yet even though the gold trade has accelerated to the hyper-speed of modern commerce, its folkways are as secretive as ever, less like a business than a court intrigue.

10

SHADOW GOLD

*When Rothschild's ran it, the daily fix took place in a
paneled room hung with portraits of departed gentry.*

ONE OF THE GOLD WORLD'S TRAITS IS SECRECY.
They polish furtiveness as if it were the family silver. Consider the
arcane practice of fixing the price.

Every weekday morning at 10:30 a dealer in the precious metals
trading room of a London bullion bank picks up the phone and
punches into a special line. The dedicated line connects him to four
other bankers. The five banks form a powerful and self-policing group
called the London Gold Fixing. Its members are: ScotiaMocatta,
the gold trading arm of the Bank of Nova Scotia; Barclays Capital;
Germany's powerful Deutsche Bank; Société Générale; and HSBC,
the London-based multinational created from the old Hong Kong
and Shanghai Banking Corporation. These banks set the price of
gold.

"There are loads of different places in the world to buy gold," said Jeremy Charles, a trim, tough-looking, fifty-seven-year-old native Londoner who started as a tea boy in the gold rooms of the legendary Rothschild bank and ended running HSBC's global bullion operations. "But at 10:30 in the morning, you find things sort of stop while the whole market watches London."

The fixing starts with the "nomination" of a price by the bank holding the rotating chair. Often this opening number is midway between the last recorded London buying and selling prices. The chairman asks who would buy and who would sell at the suggested figure. "That's communicated by each of five members to their colleagues in the dealing rooms. So let's say today's fixing starts at such and such a price, and our dealers are in touch with all of our clients, and our clients are in touch with their clients, and so on and so forth until you get right down to every man, jack, and dog that wants to follow the gold fixing."

If the numbers of buyers and sellers fail to match at the suggested price, the figure is adjusted up or down until it hits a point where buyers and sellers are equally enticed into the market with equal volumes of gold, and agree on price. "All the nominations are done in bars," said Charles. "I might say, 'I'll buy eighty bars,' and somebody else wants eighty, and if the price is agreed, it is fixed."

When the banks locate that point of equilibrium—buyers matching sellers—the "fixed" price is flashed out around the world. They repeat the process in the afternoon.

Some of the firms dealing gold in London have been doing it for centuries. ScotiaMocatta can trace its origins to 1671, when Moses Mocatta began trading bullion in London. When Jeremy Charles started in the tea room at Rothschild's in 1975, the bank still operated from its old stone building at One King William Street. The daily gold fixing took place in a paneled room hung with portraits of departed gentry. In those days, a small Union Jack flag mounted on a pivot lay on its side on the desk in front of each banker. As the fixing got under way, bankers would raise their flags and shout "Flag!" when they wanted to draw the chairman's attention to a change in their position. The price was not fixed until all the flags lay on their sides again, signaling that sellers matched buyers. Today, much of the action has moved downriver to the modern towers of the Canary Wharf financial district. Yet they still bark "Flag!" into the phone to signal new positions, and the standard of the bullion that they trade hasn't changed. Market makers deal exclusively in London Good Delivery bars. When central banks and hedge fund billionaires want gold, that's the kind they want.

No outsider is allowed to witness the fixing. When I spoke to Charles, we had our interview in a meeting room on a high floor at the bank, with a view of the Thames flowing past Greenwich. As he showed me out, I asked if we could stop by the precious metals trading room, so I could see what it looked like. "Oh, that's strictly forbidden," he said.

There's a lot about gold that is generally hidden, including how hard it is for ordinary people to deal in. The idea that gold is universal money is laughable. Who would accept it? It would have to be someone with testing equipment at his elbow. It's not easy to sell gold. That's why those who trade on the London market don't really deal principal to principal, but through bullion banks. Gold is easy to convert to cash if you work at HSBC, otherwise, not so much. A group of London criminals discovered this painful truth thirty years ago when they suddenly found themselves with a large amount of bullion on their hands, and began a tale of woe that has not stopped to this day.

· ● ·

IT WAS SUPPOSED TO BE a lark. "Mad Mickey" McAvoy, a young London criminal, and a veteran armed robber named Brian Robinson, known as "the Colonel," had heard that £3 million in cash was about to be shipped through the Brink's-Mat Ltd. security company's warehouse at Heathrow airport. They put out the word that they were looking for reliable help, assembled a team, and at six o'clock in the morning of November 26, 1983, drove out of the still dark city to the airport.

The caper was supposed to be a cinch because the robbers had an advantage not generally available to thieves—a key to the depot. They had got it from Robinson's brother-in-law, who worked there. He had also given them the general layout of the place, and told them what security measures to expect inside. They had a clear plan for circumventing those measures, namely: they disabled the alarms, rounded up the guards who had the combinations to the vault, poured gasoline on them, and threatened to ignite them unless they unlocked the heavy steel door. The guards unlocked it. The gang charged in to get the cash and discovered, instead, a stack of 6,800 gold ingots. That's when things stopped being a lark.

First they had to find a vehicle big enough to carry off the gold. They put out a call and managed to get a van. A job that was supposed to have been in and out in minutes took two hours. Then, what were they going to do with the loot? Gold is not money. If they had found the stash of banknotes they'd expected, they would have stuffed it into gym bags, gone home, and divvied it up in the basement. The only skill required was counting. But gold's conversion into money demands a different skill set, which the thieves did not possess. You could say that, right away, they found themselves facing the liquidity challenge of a precious metal. A thing is liquid if there is a ready market for it, as there is for gold. But it's not a market that you and your mates can drop your gold at on the way back from the depot. The robbers went to Brian Perry, a London hood. Perry brought in Kenneth Noye.

With the entry of Noye, a human grenade drops into the Brink's-Mat story. According to Murderpedia.org, a website devoted to murderers, Noye's favorite weapon was a knife. The *Telegraph* called him a "multimillionaire gangland killer." He had a ten-bedroom mansion in Kent, a villa in Cyprus, and a £700,000 yacht. He also had a knack for smelting, and started combining the gold with copper to create a low-grade alloy that (so his thinking ran) he could sell into the gold scrap market without attracting attention. The scrap market is the cash-for-gold business that recycles jewelry back into bullion. But it takes a lot of wedding rings to add up to £10 million, the sum that Noye and a confederate deposited, wad by wad, into a bank in Bristol. The deposit attracted the attention of bank inspectors, who tipped off Scotland Yard.

The police set up surveillance on Noye. On a January night in 1985 Detective Constable John Fordham was in Noye's backyard watching the house when Noye's Rottweiler dogs spotted him. British policemen do not regularly carry guns, and Fordham was in camouflage. Noye attacked him with a knife, stabbing him eleven times and killing him. At his murder trial Noye convinced the jury that he had believed himself in danger from a prowler. He was acquitted. But investigators rooting through Noye's property found eleven ingots wrapped in cloth and buried in a drainage ditch. Noye went to prison for conspiracy to handle stolen goods.

The original perps, the Colonel and Mad Mickey, were already in jail. The police had quickly sussed out the gang's security guard insider, who had confessed. Thirteen months after the robbery the two ringleaders were tried at the Old Bailey, and sentenced to twenty-five years apiece. Apparently they tried to strike a deal for lighter sentences by revealing the location of the gold. But the gold was no longer where they thought it was, because Perry had already passed it on to Noye. Or anyway, some of it.

Criminals lead chaotic lives whether they're stealing women's purses or gold bullion. Noye got out of jail in 1993. In 1996 he stabbed a motorist to death in a road rage incident. He escaped to Spain in a jet supplied by John "Goldfinger" Palmer, a Bristol gold dealer and fraudster who'd been charged with handling Brink's-Mat gold, but acquitted. Two years after Noye's escape, police tracked him down and brought him back to England. In 2000 he was sentenced to life imprisonment. A year later Brian Perry, who had enlisted Noye, died in the street when assassins shot him three times in the back of the head. Other bodies too were crumpling to the floor. In 2012 the *Mirror* newspaper listed eighteen men who'd been killed over the years in connection with the Brink's-Mat gold, including an underworld enforcer thought to be entombed somewhere in the concrete foundations of the O2 concert arena in Greenwich, southeast London.

The stage is heaped with corpses, but where's the gold? Some vanished in the Bristol scrap sales, some was recovered from Noye's ditch, and in 2008, in a raid on thousands of safety deposit boxes across London, police discovered luggage stuffed with plastic parcels of gold grains. "They were wrapped like large packets of peas," a Scotland Yard commander said. The raid was an unrelated operation to uncover criminal proceeds, but detectives speculated that the gold was part of the Brink's-Mat hoard. About two thirds of the original haul is still missing, and today it would be worth half a billion dollars. In 2012 a detective told the *Daily Mail* that he thought the rest of it was still around, and that at least two London gangs had started "breaking a few arms" to find it.

GOLD AND CRIME GO HAND in hand. One of the sublime rituals of money, the Trial of the Pyx, is aimed at catching cheats. In a ceremony dating from 1282, porters from the Royal Mint lug chests of coins to the Goldsmiths' Hall where a red-robed English judge called the Queen's Remembrancer waits with a jury of assayers. But that is the easy part of policing gold. The schemes that spook the gold trade can't be sorted out so simply. They are the shadowy events, the phantoms rippling through the tall grass of the business, not clearly visible or understood, and sowing panic for that very reason.

One such passage shook the market for two weeks in July 2010, when a massive, unexplained transaction parked a load of bullion in a place where those who monitor such movements did not expect to find it. When they discovered it, their confusion about why it was there was deepened by the stubborn silence of a little known Swiss bank, the Bank for International Settlements, or BIS. Headquartered in Basel, BIS is sometimes called the central bankers' bank, because that's where they go to borrow. For that reason, its power is very great. Federal Reserve chairman Ben Bernanke sits on the board, with such other central bankers as those of Britain, France, Germany, Canada, and Japan. I can tell you from experience, they have the tightest lips in Europe. BIS has raised the game of saying nothing to the level of Olympic sport. In any contest where the aim is to stonewall, they would sweep the medals. Their failure to provide a full account of the transaction added yet another layer of opacity to an already secretive affair, and down went the gold price in a fright.

The cause of the panic was a footnote to the bank's annual report. The note revealed that a bank or group of banks had lent 349 metric tons of gold to BIS in exchange for cash. The deal was so colossal—a sixth of the world's annual production—that the news of it, without elucidation, stunned the bullion market. Who had lent the gold, and why? Was a bank in trouble? Could it be a central bank? What did they know that the bullion market didn't? As questions thickened the air, investors started bailing out of gold, and the price lost $40 in a day.

Here's why the price went down. The deal looked like a swap—an exchange of gold for cash, with an agreement that the gold would be redeemed at a later date. Like taking jewelry to the pawn shop, you get it back when you repay the loan. For whoever swapped it, the gold raised $14 billion. What bothered the gold market was the question of what would happen to the gold if the $14 billion didn't get paid back. The price was down anyway, depressed by a period of selling by ETFs. Would the 349 tons suddenly appear for sale if the swapper couldn't manage the loan? A 380-ton gold dump would annihilate the price.

Suspicion swirled about who might need the cash. There were plenty of candidates. Fear of European countries reneging on their debts was in the air already, and this fear suggested an explanation for the swap. A central bank desperately needed cash.

For a while that's what everyone I spoke to thought—Greece or Spain or some other monetary basket case was being hooked up to life support. But then Edel Tully, an analyst at UBS, pointed out that it couldn't be a central bank. European regulations, she said, did not allow a central bank to transfer funds to its government or to buy the government's bonds—the actions they would have to perform to stave off a default. Not a central bank, then, but some other monetary authority. On her list of suspects, Tully placed the International Monetary Fund. Suspicion switched to the IMF.

The fund is the world's banker of last resort. It had bailed out Iceland, the first sovereign meltdown of the financial crisis. The IMF had been "quietly selling off its gold" position anyway, according to the *Telegraph*. In a piece headlined "Secret Gold Swap Has Spooked the Market," the newspaper said that the IMF was chronically short of cash. No sooner did opinion settle on the IMF, than the BIS itself torpedoed it. It was not the IMF, or even a central bank, that had swapped the bullion, they said in Basel, but a commercial bank or banks. If this announcement was supposed to calm the market, it failed. The 349 tons of gold was far too much for the market to believe that a single commercial bank had accumulated it. The sheer size of the transaction—*The Wall Street Journal* said that it tripled the amount of gold the BIS had owned the month before—left the market bewildered. Even for a cartel of banks it was a lot of gold, and the question still remained— why? Whose cupboard was so bare they had to pawn the family china?

"It's odd, but it could be bad," said a senior metals strategist in London, succinctly capturing the bullion market's mix of bafflement and fear.

As the speculation crackled on, it attracted the Gold Anti-Trust Action Committee (GATA), a group that believes there is a broad conspiracy to manipulate the gold price. "As always, news of anything to do with the gold market is cloaked in secrecy, misinformation, and innuendo," an article on GATA's website said. It called the swap a "tripartite transaction" in which a "commercial bank or banks made a swap with a central bank or banks and then the commercial bank or banks made a swap with the BIS." I diagrammed this with little boxes and arrows, and stared at it until my head hurt. I still don't get it, but GATA and the BIS have history.

In 2000 a GATA-funded litigator sued the bank, the U.S. treasury secretary, the chairman of the Federal Reserve, and others in the U.S. District Court for Massachusetts. He accused them of "price fixing, securities fraud, and breach of fiduciary duty," and with exceeding their lawful authority. The plaintiff was Reginald Howe, a lawyer and investor who had shares in BIS. BIS had decided to liquidate all such private holdings, and Howe alleged that the price they were offering undervalued the bank's bullion holdings as part of a deliberate plan to suppress the gold price. The government's motive in suppressing price was to conceal an indicator of the American inflation rate. Since gold was denominated in dollars, a rising gold price would reveal the true state of the dollar. Although Howe failed in his larger purposes, a European tribunal ruled that he and others had been underpaid by BIS for their shares, and ordered the bank to pay them more.

But back to the swap. Chilled by the steady drizzle of apprehension, the gold price had fallen $80 when, at the end of the month, the *Financial Times* announced that the mystery had been solved. The swap, the paper said, had been a BIS idea all along. BIS had been looking for a way to make money out of its large dollar holdings, and suggested the gold swap to some European commercial banks that happened to want dollars. More than ten European banks agreed to swap their gold for cash. "From time to time," a banker said, "the BIS want[s] to optimise the return on their currency holdings." The swaps, such sources told the *FT*, had been mutually beneficial.

Let's look at that. If the swaps had been a purely benign activity, why did it take so long to get an explanation from the BIS? When bullion goes tiptoeing through Europe to a secretive Swiss bank controlled by foreign treasury officials, it's fair to ask if there's an explanation other than business as usual. There is.

After the 2008 banking crisis, the prospect of mass bank failures prompted European regulations. The regulators set new capital requirements for commercial banks. Many of these banks were holding too many bonds of countries whose credit ratings had gone bad. To see if the commercial banks were complying with the new standards, the European Central Bank planned to conduct audits called "stress tests." A reasonable suspicion about the gold-for-dollars swaps, then, was that it was a subterfuge for packing cash into the European banking system to improve the balance sheets of private banks in advance of the stress tests. The commercial banks were so shaky that their own central bankers had become reluctant to deposit money with them, placing their extra funds in BIS instead, hence all the extra cash. The swap gave the international treasury establishment a way to avoid the crisis that would certainly have followed stress test failures. The only dupes would be the public . . . *and the people who actually owned the gold.* The swapped gold did not belong to the commercial banks. The banks had borrowed it, mostly from their own customers. If their customers had asked for it, the gold would not have been there.

· · ·

THE COMMERCIAL BANKS HAD OBTAINED the gold from two sources. Some of it they'd borrowed from central banks. The other source, their own customers' gold, would have come from "unallocated" accounts. The gold in an unallocated account, unlike that in an allocated one, is not held separately for the owner. The bank can trade it until you ask for it back.

In the swap with BIS, the commercial banks were getting cash for other people's gold, and using the cash to look healthier than they were. This maneuvering was not illegal, but perhaps more creative than we like a bank to be. The bank had used the gold the way it used cash deposits. They "owed" the gold to the owner in the same way that all deposits are obligations. But if a bank was in such straits that the BIS seized the pledged gold, you wouldn't like the bullion owner's chances of recovering his metal. And his title might be tangled anyway.

The depositor himself might already have borrowed against the gold while keeping ownership. In such a case the gold is said to be "hypothecated"—ownership not transferred, but the asset hypothetically controlled by someone else with the right to seize it: the bank. If the bank then borrowed against this hypothecated gold by swapping it for dollars, the gold would be said to have been "rehypothecated." The gold that was swapped for cash went off to Basel with a tail of obligations flapping out behind.

Since part of gold's allure as an investment is its supposed safe-haven function, it's fair to wonder what would happen if suddenly those who owned paper claims wanted the actual metal in their hands. In any climate of doubt about the amount of physical gold backing up paper gold, investors will think about ETFs. Do the funds have the gold they are supposed to have? If they have allocated gold, then presumably it is stacked up in its own dedicated corner of the vault. As long as the hoard has not been compromised by some of the new tungsten-alloy fakes making their way into the market, the investor is covered.*

A more realistic worry for the investor is what would happen to gold ETFs in the case of a panic. Say a free fall in the stock market prompts margin calls, and some investors decide to sell gold to cover their positions. Let's imagine that a lot of investors bailing out of gold own small pieces of the same ETF. Taken together, the small pieces are a big piece. When the investors try to redeem their stakes, there's not enough gold on hand. The ETF evaporates in a puff of insolvency. The failure sparks redemption calls on other ETFs, ones that actually do have physical gold to back up all shares. These sturdier ETFs start selling gold to meet redemptions. A cascade of bullion splashes into the market. The gold price tanks, and so do gold mine stocks. Galvanized by this collapse, the run on equities turns into a rout.

Even if you find this scenario farfetched, it's still what gold is supposed to save you from. Part of gold's investment raison d'être is just such a possibility: for peace of mind, buy gold. But gold is never peaceful. It's a fever spread by doubt. Doubt and suspicion are its pathogens. Besides jewelry and a few industrial uses, there are no other reasons to own it. If you're not suffering from the fever, you are betting that others will. Gold is inseparable from speculation about

---

* The technology of faking gold is getting better, most of it apparently based on the near-identical densities of gold and tungsten. The counterfeiter drills out plugs of gold from a bullion bar, replaces the gold with an alloy of tungsten, and covers the replacement plug with a thin layer of gold. The *Financial Times* reported a Chinese scam that involved "a complex alloy with similar properties to gold." The fake contained about 51 percent bullion, and seven other metals: osmium, iridium, ruthenium, copper, nickel, iron, and rhodium. A skilled metallurgist had made it. One of Hong Kong's biggest jewelers, the Luk Fook Group, was fooled by such fake gold.

disaster, misfeasance, or manipulation. Even when the price is high, owners suffer from the fever, because what if the price goes down?

AT THE MORNING FIX IN London on September 5, 2011, gold hit $1,896.50 an ounce. Propelled by a ten-year bull run, the rising price had an air of inevitability about it. The hurdle of $2,000 was suddenly right there—a cinch! For a year, analysts had been predicting gold would clear the bar. Like everyone who followed gold, I wondered how long it could continue, and how the price could be justified. By luck, I had another appointment with Peter Munk. His office had contacted me to say that he was passing through London. He wanted to talk about price, and we were to meet just as gold was pushing up against $2,000.

A gale was chewing its way through southern England as I got a taxi and went down to Mayfair. In Park Lane the wind was blowing out umbrellas and thrashing people's coats against their legs. As we passed the Dorchester Hotel, a geranium with a clod of earth still clutched in its roots sailed into the traffic and exploded on a car. In the Four Seasons lobby, women were stealing glances at each other's hair and shaking water from their shoes. Promptly at 5:15 Munk popped from the elevator. He scowled at the crowded lobby, seized my arm, and said, "We'd better go upstairs."

He wore a dark blazer and charcoal pants and a soft-collared, light blue shirt. The white enamel pin of the Order of Canada glowed on his lapel. Munk had just been to the Venice Biennale with his wife. In Venice they had stayed on his 140-foot yacht, *Golden Eagle*. It takes a crew of nine to sail it. "You think that's big?" said Munk. "That's nothing! They made me moor at the farthest point of the basin to make way for people with much, much bigger boats. They look like cruise ships. You have no idea how big they are!"

I did have an idea, because I had been reading up on them in articles about Munk's latest venture, a marina for super-yachts at Tivat, the former Yugoslavian naval base in Montenegro. Munk used the profusion of these gigantic pleasure tubs to illustrate his theme—that there were more rich people in the world today than there had ever been, and they had assets to protect.

"Today I talk to people, and I have been around for a long time and I am associated with gold, and they confide in me. And everybody I speak to is now buying gold. They are not gold bugs, Matthew, and you know that Peter Munk is not a gold bug either." He stood up from his seat and looked out the window at the rain with a fierce expression. Gold bugs are by definition fanatics—believers in gold as the only real money. I knew that Munk subscribed to no such theory, although I guess he had no problem selling gold to those who did. His point, though, was that a new supply of gold buyers had appeared on the rising tide of recently created wealth.

"These people I talk to," he resumed, "they buy because they want protection to keep a portion of their assets and they think gold will do that. They think the [financial] system is not going to change unless there's a catastrophe, a French Revolution, a war, or a devaluation of such proportion that the United States government defaults. These people who have been accumulating gold are not going to be selling."

Yet they were. A Zurich-based executive of Stonehage, a firm that provides wealth management advice, had been telling Reuters only days before that some rich clients had already taken what they'd made in gold's decade-long bull run and put it into what they saw as the next smart thing, high-end art. This shift accounted for substantial gold amounts, because some of the clients had had as much as 15 percent of their assets in bullion. The money that had gotten into gold ahead of the pack was also getting out ahead of it, and just in the nick of time. The day after I spoke to Munk the gold price sank. In three weeks it dropped $300.

Among those battered in the fall was John Paulson, the American hedge fund sorcerer. Paulson made staggering sums on gold. In 2010 he personally made $4.9 billion on his gold investments. The *New York Times* calculated that this paycheck came to twice the total player salaries of Major League Baseball. He lives in a 28,500-square-foot limestone mansion on Manhattan's Upper East Side. Most of Paulson's own gold money was invested in a special gold-share product that his firm sold. The shares did not represent gold bullion, but were linked to gold-based assets such as gold mining stocks.

Paulson was the man with the Midas touch, until he wasn't. His fortunes started to turn with the collapse of an $834 million investment in Sino-Forest Corporation, a Toronto-listed Chinese timber company with property in Yunnan province. He had been accumulating stock since 2008. By 2011 he owned a majority stake. In June that year a firm of short sellers called Muddy Waters accused Sino-Forest of being a classic "pump and dump," an operation in which the owners of a stock use false information to inflate its value— the pump—and later sell when the share price is high—the dump. Muddy Waters claimed Sino-Forest had overstated timber reserves by $900 million, and had run what amounted to a twenty-year fraud. The stock collapsed. Paulson took a loss of half a billion dollars. One of his largest funds lost 47 percent of its value. That was around the time that the gold price got on the elevator and pressed the down button. Paulson's gold fund lost 16 percent in a single month.

One report said Paulson was so angered by the steady leakage of news about his losses that he tightened his reporting policies to make it harder for the media to get information. The only public documents recording his gold actions were the quarterly 13F filings mandated by the SEC. The 13Fs showed Paulson dumping gold. He sold a third of his company's position in the Spider for $1.4 billion. According to an unnamed source cited by Reuters, Paulson's move from the Spider was not an abandonment of gold, but a switch to forms of ownership that did not have to be reported in public filings, such as swaps, forwards, and physical gold. He was moving his gold from sight.

A STORE OF GOLD IS a hiding place. The Oxford dictionary defines "hoard" that way, as something "hidden away or laid by." But laid by against what? Gold is a doubtful sanctuary. As the price was tanking that September, a financial reporter attributed the fall to investor preference for stocks, because the stock market was rising in response to a rare moment of good news about the Greek debt crisis. Three weeks later the debt crisis returned to its usual state of hopelessness, and equities fell. So did gold. This time the reporter said that the falling market had dragged the metal down. Market up: bad for gold; market down: bad for gold.

Why is gold worth anything at all? Those who buy it perform a rite so old it's scarcely possible to separate it from who we are. It's the archaeologist's dilemma: we are inside the object we are trying to understand. In *The Golden Constant*, a classic study that tracked gold's purchasing power through time, the economist Roy Jastram confessed to a "nagging feeling that something deeper than conscious thought, not an instinct but perhaps a race-memory," was behind our attachment to gold.

Among the uses for gold listed on its website, Barrick includes such devices as medical thermometers capable of detecting the tiniest changes in a patient's temperature. And it's true that gold makes a good thermometer, but not because it measures what is happening to the human body. It's good because it measures what is happening to the human race.

11

THE GOLD IN THE
BAMBOO FOREST

Empires rose and fell in the desert—
secretive, enigmatic, fabulously rich.

IN *KING SOLOMON'S MINES*, THE VICTORIAN POT-
boiler whose adventurers find treasure in the heart of Africa, a book
captured the spirit of an age. It was a publishing sensation. Its author, H. Rider Haggard, became a millionaire. The reading public,
captivated by such recent discoveries as Egypt's Valley of the Kings,
snapped up the novel so fast that the publishers had trouble keeping it in print. When Alan Quatermain, the swashbuckling hero,
heads up an ancient road to the royal city of Loo, the reader found
the exploit believable. Not long before, explorers had discovered
the ruins of Great Zimbabwe, a stone city whose oldest buildings
dated from the eleventh century. Haggard's book created a new lit-
erary genre, inspiring such bestsellers as Edgar Rice Burroughs's *The
Land That Time Forgot* and Arthur Conan Doyle's *The Lost World*.

The success of these books revealed a whole country's hunger for adventure.

Today the gold rush is that adventure. It catapults us out of the quotidian into an enticing dream-world of riches and romance. Africa brims with gold. We emerged from Africa. Maybe the idea of gold came with us.

I thought of this one night on the balcony of a hotel in Dakar, when I couldn't sleep. In the morning I was leaving on a journey across Senegal to a distant goldfield. I would cross the territory of vanished empires as fantastic as any that Haggard had imagined. The gold rush has awakened them. The forests teem with people whose gold mining skills originated in prehistory. The gold empires of Africa had captivated Europeans centuries before Haggard wrote his book. Portuguese ships had gone scraping past Dakar in the 1400s as they probed the coast of Africa for the gold kingdoms. They had not found them.

I gave up on sleep and phoned for coffee. There was no moon. The electricity had failed in the night. The ocean lay a hundred feet away, invisible, wrinkling softly on the beach. The desert deposited a layer of grit on the balcony, grainy underfoot. We left Dakar at 5:00 A.M.

A succession of empires rose and fell in West Africa from about the seventh century, when a ruler called Dingha Cissé established the Wagadu Empire. It was a secretive and enigmatic power, fabulously rich, in the Mauritanian and Mali deserts. The Soninké people domesticated camels and established trade routes to North Africa. They traded salt, slaves, and gold. The location of the mines was a state secret. The monarch could field an army of 200,000, including 40,000 archers and a strong cavalry. A merchant who visited the ruler's court in the eleventh century gave this account:

He sits in audience or to hear grievances against officials in a domed pavilion around which stand ten horses covered with gold-embroidered materials. Behind the king stand ten pages holding shields and swords decorated with gold, and on his right are the sons of the kings of his country wearing splendid garments and their hair plaited with gold. The governor of the city sits on the ground before the king and around him are ministers seated likewise. At the door of the pavilion are dogs of excellent pedigree that hardly ever leave the place where the king is, guarding him. Around their necks they wear collars of gold and silver studded with a number of balls of the same metals.

In the myths of the Soninké, every year a seven-headed snake called Bida replenished the gold in the mines. In exchange for the annual sacrifice of a maiden, Bida caused a rain of gold. This arrangement ended when a young man decided to keep the maiden for himself, and cut off Bida's heads. The gold rain stopped. The mines ran out. The Soninké lost their empire.

Over the centuries one empire melted into another through decline and conquest. News of these kingdoms reached Europe, and the Portuguese went looking for them. They established a trading port, but failed to find the gold source. In 1698 the Dutch traveler William Bosman described what the adventurers believed about the people who lived inland. "They are possessed of vast treasures of Gold besides what their own Mines supply them with, by plunder or their own commerce."

The Ashanti state was the last of the African gold empires. In its founding story, a golden stool descended from heaven into the lap of the paramount chief Osei Tutu I. The federation started by this leader in the seventeenth century eventually extended into the Sahara, absorbing parts of the old Wagadu Empire. The Ashanti king, called the Asantehene, ruled a population of about 3 million. Extravagant reports about him circulated. A Danish doctor wrote that "this mighty king has a piece of gold, as a charm, more than four men can carry; and innumerable slaves are constantly at work for him in the mountains, each of whom must collect or produce two ounces of gold per diem." The Ashanti state had a treasury filled with gold that was cast into standardized weights. They traded gold at outposts on the Atlantic, a trade that gave the country its European name—Gold Coast. Inevitably, the traders wanted to see for themselves where all the gold was coming from. On April 22, 1817, the British consul Sir Thomas Bowdich marched inland to find the Ashanti capital.

They set off in good order. A breeze came off the sea. They entered the green shade of the jungle on a pathway paved with pulverized quartz. Then the jungle sucked the breeze away, then the sunlight. The men advanced into a steamy, twilit furnace of vegetation. The quartz path ended. Then there was no path at all. They struggled forward into mangrove swamps. Their nostrils filled with the stench of rotting vegetation. Sweat poured into their eyes and soaked their clothes. Bowdich wrote:

> *The ground of our resting place was very damp, and swarmed with reptiles and insects; we had great difficulty in keeping up our fires, which we were the more anxious to do after a visit from a panther. An animal which, the natives say, resembles a small pig, and inhabits the trees, continued a shrill screeching through the night; and occasionally a wild hog bounced by, snorting through the forest, as if closely pursued.*

They marched for two weeks and came to the ruined villages of the Fante, a people crushed by the Ashanti. Ashes and skulls littered the township. The troop set off again and crossed the Pra River into Ashanti country and found clean villages with wide main streets. On May 19, 1817, they halted a mile south of Kumasi and changed into scarlet uniforms and sent messengers ahead to announce their arrival to the Asantehene, who certainly knew of it already. He sent word for them to wait until he finished bathing. When the king was ready, messengers told the British to enter the city. They marched in at two o'clock in the afternoon, passing under a suspended fetish, a dead sheep wrapped in red silk.

Thousands of people packed the road to stare at the first Europeans most of them had seen. Massed Ashanti warriors filled the air with a shattering din—horns, drums, rattles, and gongs. Fusillades of musketry rolled a dense curtain of smoke across the visitors. So thick was the smoke that the British could only see the path immediately in front of them until they reached a clearing in the crowd. In the open space, flag bearers sprang from side to side waving banners, and the captains leapt in to dance.

The Ashanti captains wore fantastic war hats with gilded rams' horns and plumes of eagle feathers. Their red cloth vests were decorated with fetishes and passages of Arabic script stitched in silver and gold. Leopards' tails hung down their backs. As they danced and vaulted in the ring, small brass bells fixed to their costumes jingled. They wore red leather boots that came up to their thighs. Quivers of poisoned arrows dangled from their wrists. Each captain gripped a length of iron chain in his teeth.

When the dance ended, the Europeans were squeezed along through narrow lanes in the packed multitude. They saw streets with long vistas jammed with people, and houses with open porches where women and children clustered to watch them pass. When they neared the palace, horns and flutes played "wild melodies," while huge umbrellas were used to stir the air and refresh the British with a breeze.

As the soldiers waited to be summoned, a troop of guards with caps made of shaggy black skin led a prisoner by on his way to execution. A knife pierced the man's cheeks and his lips were sewn shut, and his body showed the wounds of other tortures. The escort pulled him along by a cord through his nose. Then the British were summoned forward.

> Our observations . . . had not prepared us for the extent and display of the scene which here burst upon us: an area of nearly a mile in circumference was crowded with magnificence and novelty. The king, his tributaries, and captains, were resplendent in the distance, surrounded by attendants of every description, fronted by a mass of warriors which seemed to make our approach impervious. The sun was reflected, with a glare scarcely more supportable than the heat, from the massy gold ornaments, which glistened in every direction.

More bands burst out—drums, flutes, and bagpipes. Noblemen and members of the royal family lined the way, and high officials of the kingdom—the gold horn blower, the chamberlain, the master of the bands. They wore brilliant clothes and massive jewelry. Gold necklaces drooped to their waists. Gold bands circled their knees, and disks and rings and little casts of animals, all made of gold, jiggled and clinked on ankle chains. The most important men had heavy nuggets hanging from their wrists, "which were so heavily laden as to be supported on the head of one of their handsomest boys."

Above the court a sea of huge umbrellas rose and fell as the bearers moved them in an undulating wave to churn the air into currents. The cloth of these giant parasols was sewn from pieces of yellow and scarlet silk. The tips blazed with gold ornaments—elephants, pelicans, crescents, and swords. The umbrellas flashed and twinkled as the sunlight played on tiny mirrors sewn in the cloth.

A huge man with a heavy gold hatchet slung across his chest stood near the king. He was the executioner. His attendant held the execution stool, thick with clotted blood. The keeper of the treasury displayed his symbols of office—solid-gold scales and weights, a blow pan and boxes. Under its own umbrella sat the golden stool, the symbol of the nation. The king's soul washers wore gold disks or golden wings. The soul washers caught any evil directed at the king, deflecting it with their gold insignia. Four linguists stood near the monarch. The Ashanti had no writing. The linguists were their living archive, with an encyclopedic knowledge of tribal lore and proverbs. They acted as spokesmen for the king, and ambassadors, and carried staffs topped by gold finials with finely wrought designs—a spider-web, an antelope with antlers full of birds.

The king sat in the center of his court, in a chair covered in gold. Attendants waved a veil of elephants' tails spangled with gold in front of him. Bowdich thought he was about thirty-eight years old. He wore a dark green cloth. A ribbon of glass beads circled his temples. A red silk cord across his shoulder held three fetishes wrapped in gold. Gold rings hid his fingers. A white crown was painted on his forehead. He had gold castanets in one hand, and could bring the court to silence with a click.

After greeting the king, the British were conducted to a tree some distance away. The whole court milled around and put itself in order for the next stage of the proceedings: repaying the visit. Now the sea of umbrellas, springing up and down in a billowing parade, advanced on the guests. Chiefs rode in crimson hammocks. They dismounted thirty yards from the British and approached to welcome them. Regiments marched past the visitors. Bowdich and his officers reckoned there were 30,000 men in military order. It was late in the evening, "a beautiful star light night," as Bowdich wrote, when the king himself approached. Torchlight glittered on the Asantehene's regalia. The skulls of enemies decorated the largest drum. The king stopped and asked the British to repeat their names, then said good night and at last retired, followed by a throng of sisters and aunts shimmering with gold.

I spent an afternoon in the British Library looking at maps the first cartographers had made of the Ashanti lands. On top of one I placed a sheet of acetate so I could flatten the paper and examine the exquisitely drawn huts and streams. The prowling lions looked like large and irritated spaniels. The whole top quarter of the map was colored in a pale green wash and annotated in a flowing script. "Rich in gold," the cartographer had written, "found in nuggets in pits nine feet from the surface. Brought to Kumasi in solid lumps embedded in loam and rock which together weigh fifteen pounds."

In 1824 Britain began the first of its campaigns to subdue the Ashanti. They sent expeditions against them from the coast, and seized Ashanti gold mines. The final conflict began when the British governor Sir Frederick Hodgson arrived at Kumasi in April 1900 and demanded the most sacred object in the land.

Where is the Gold Stool? Why am I not sitting on the Golden Stool at this moment? I am the representative of the paramount power; why have you relegated me to this chair?

The Ashanti seem to have been dumbfounded by the deadly insult. But that night, in a secret meeting, the queen mother Yaa Asantewaa poured a scalding speech onto the men.

> *Now I have seen that some of you fear to go forward to fight for our king. If it were in the brave days, the days of Osei Tutu, Okomfo Anokye, and Opoku Ware, chiefs would not sit down to see their king taken away without firing a shot. No white man could have dared to speak to a chief of the Ashanti in the way the Governor spoke to you chiefs this morning. Is it true that the bravery of the Ashanti is no more? I can't believe it. I must say this, if you, the men of Ashanti, will not go forward, then we will. We, the women, will. I shall call upon my fellow women. We will fight the white men. We will fight till the last of us falls in the battlefields.*

The queen's words stirred the men into a rage. In the events that followed, sometimes called the Yaa Asantewaa War, a British detachment, unaware of the mounting danger, went hunting in the nearby bush for the golden stool. The hostile Ashanti engulfed them. Only a sudden downpour saved the soldiers, covering their retreat. A six-month war ended in Ashanti defeat. The British exiled the queen to the Seychelles, where she died. The Ashanti lands were merged into the Gold Coast, now Ghana.

I DROVE OUT OF DAKAR with Martin Pawlitschek, a forty-three-year-old Australian geologist who lived in the city with his wife and two children. Tall, with pale blue eyes and light brown hair, Pawlitschek has an easy, affable manner. We had a thermos of coffee and a package of almond tarts. I felt the keen pleasure of being up before dawn. We blundered through the dark streets until he found the highway he was looking for. Masses of people crowded the dark edge of the road. Women with baskets on their heads swayed along in ankle-length skirts. Merchants in white galabias took the shutters from their shops, and roadside stalls bloomed like predawn flowers. Riders on Chinese motorcycles shot through the traffic, weaving among the overloaded trucks that tilted as they dodged the potholes.

We were setting out to drive across Senegal to a package of gold targets in the hills along the Mali border. Geologists had known for years that the ground was promising, and when Senegal passed a new mining code in 2003 that protected investors, in came the drills. An Australian company struck gold, developed a mine, then asked a banker in Toronto to find someone to run it. He found Alan Hill.

Hill had been running a Romanian gold mine, but had quit in the face of "frustrations." Romania had been a gold producer from antiquity, although in 2000 it became better known for producing catastrophe, when the tailings pond of a mine in the ancient gold mining center of Baia Mare ruptured, spilling 3.5 million cubic feet of cyanide into a tributary of the Danube.

When Hill left Romania, his Canadian management team came with him. The Australians hired them all and formed Teranga Gold Corporation. (Teranga means "hospitality" in Senegal's dominant Wolof dialect.) Teranga was floated in Toronto. It raised $145 million. The main assets were the company's 130,000-ounce-a-year Sabodala gold mine located on the original discovery, and a glittering package of exploration targets. Hill planned to double the size of the mill, and Pawlitschek's job was to find the gold to feed it.

· ● ·

BY THE TIME THE SUN came up we were clear of the city. A cloud of flamingos descended on the Saloum salt flats. The chimney of a salt mill belched dark smoke. Solitary baobob trees scratched at the sky with their demented branches. We came to a stretch of highway pocked with craters. A semi had put a wheel into one and lay on its side like a shot rhinoceros. Boxes and packages had broken loose from the trailer and spilled into the shrubbery. Guards crouched beside the fallen cab.

Driving east across Senegal is a journey backward into time. The accidents of modern life peel away. Traffic peters out, leaving the road to the long-range trucks that ply the route to Mali. The plain, dotted with thirsty trees, extends to the horizon. Thin cattle search the brown grass. Horse-drawn carts appear on tracks beside the highway. Cinder block houses give way to the thatch and mud-brick of the villages.

At Tambacounda we stopped for lunch at a roadside hotel. The grounds were thick with neem trees, a species of mahogany imported from India by the French, who planted them in the villages for shade and because they are supposed to keep mosquitoes away. We parked beside another white Land Cruiser. Jean Kaisin, a Belgian geologist living in Dakar and hired by Pawlitschek, was headed for the gold camp with two Senegalese geologists. Also riding with him was an elfin woman from Dakar named Awa Ba, who drove an ore truck at the Sabodala mine. She wore a fuchsia-colored tracksuit with a blaze of silver sequins. She sat quietly while the men talked around her. The waiter brought us steaks as tough as planks. As the men sawed and struggled around her, I watched the cutlery flash in her delicate hands. We were all still hopelessly adrift in the task when she laid her knife and fork neatly on her empty plate.

After lunch we headed off in convoy. The vegetation grew more sparse. Thorn trees dotted the desert. Where villages clustered at the road, women in stalls sold a drink called *thé bouye,* made from monkey bread, the dry fruit of the baobab tree. When we stopped for gas, boys came to beg for coins. I learned later that they were pupils from a madrassa, an Islamic school, and had been sent out by their teachers, a practice the local people disapproved.

Late in the afternoon we crossed the Gambia River. In the shallow waters below the bridge, people panned for gold. We entered the Mako Hills. We climbed into a rocky forest. A gray pig the size of a Fiat pranced out of the bush and crossed the road without a glance, melting into the trees. The rock was a rusty pink. Pawlitschek said that it had oxidized in the open air. The resulting color suggested the presence of iron carbonate. Such a rock would have come to the surface in a hydrothermal flow of the kind that transports gold. The Mako Hills were part of a geological feature called the Kédougou-Kéniéba Inlier. We had driven onto it when we crossed the river.

An inlier is a window of younger rock pushed up into older rock. The Kédougou-Kéniéba Inlier was composed of 1.6-billion-year-old rocks. About 40 percent of the inlier lay in Mali and 60 percent in Senegal. In Mali, three large gold mines fattened their balance sheets on the formation. The Senegalese side had remained relatively unexplored. Pawlitschek was eager to show that what had been found in Mali would be found, in the same rocks, in Senegal.

Shadows lengthened on the road. We would not reach the camp in daylight. Twelve hours after setting out we broke our journey at Kédougou town, at a small hotel on a cliff above the Gambia River.

I dug out some photocopies of old maps and went to find a place to spread them out.

After the dusty road the hotel was an oasis. Guests stayed in thatched cabins in a palm plantation. A cool breeze rattled the fronds. White parrots shuffled on their perches in an aviary, and four crocodiles dozed in a heap inside a cage. Two American girls at the pool flashed their Ray-Bans at me before returning to the study of their toes. I settled into a chaise and leafed through images of the Mali Empire.

Of the gold kingdoms that rose and fell in the desert, the empire of Mali was the one that haunted the search for gold in Senegal. Its founder was a prince called Sundiata Keita, "hungering lion." The empire lasted from about 1230 to 1600. The exploration ground that we were headed for had supported gold mines that contributed to the wealth and power of this dominion. The Mali Empire was unknown to Europeans until the appearance of its greatest ruler, Mansa Musa.

Mansa Musa built mosques and palaces in cities such as Timbuktu and Djenné. In the imperial capital Niani he constructed an audience hall with an enormous dome. One tier of windows was framed in silver foil and another in beaten gold. At its height in the twelfth century, the empire comprised hundreds of cities and towns. A large urban population lived along the Niger River. Gold mines produced the kingdom's wealth. Sometimes called Lord of the Mines, Mansa Musa captivated the European imagination. In the Catalan Atlas of 1380, almost the whole of Africa is blank. In the center of the map, instead of countries and rivers, a black king in gold regalia sits on a golden throne. In one hand he brandishes a fist-sized nugget. When the Portuguese had sailed down Africa looking for a port, it was Mansa Musa's kingdom they were looking for. And so were we.

I met the geologists for dinner in an open dining room behind an oleander hedge. The sun sank and the river glowed like copper. Far away across the bush rose the purple mass of the Guinea Plateau. Below us a boatman drifted down the current. Conversation turned to the Malinke people, who panned for gold in the river as they had for centuries. They mined gold throughout the region, as they had in Mansa Musa's time. Jean Kaisin, who spent years in that part of Africa searching for the emperor's mines, told us the story of Mansa Musa's great journey.

In 1324 the emperor set out to make his hajj. He had a retinue of 60,000 soldiers and retainers and 12,000 slaves. Heralds in silk livery carried gold staffs and proclaimed the king. He had a treasure train of eighty camels, each with a load of gold dust. It's probable that so much wealth had never been assembled into one cargo in all of history. The king brought it to give away.

Friday is the Muslim holy day, and every Friday of his journey, no matter where he was, Mansa Musa paid for the construction of a mosque. In Cairo he made so many lavish gifts that he flooded the gold bazaar, and the price collapsed. A single man disrupted the Mediterranean gold market—Europe's market. An obscure, little known desert kingdom broke into European consciousness as a land of immeasurable wealth. By the time the Mali Empire passed, Europe's gold obsession was chewing up other kingdoms. The desert mines seemed to disappear from history. Mansa Musa's fabulous deposits lay largely forgotten until one day in 1989, when an explorer poking on a hill found an abandoned gallery a quarter of a mile long.

THE MAN WHO FOUND THE emperor's mine wasn't a geologist, or even a miner, but a character whose story seems taken from fiction. He stumbled on a clue and seized it, and uncovered a lost treasure.

Mark Nathanson was the son of a wholesale grocer from Cape Breton Island, Nova Scotia. He married into a wealthy Taiwanese business family. In the 1980s, among other commercial travels, he began to visit Mali. The country was then a Soviet client state, but Nathanson, learning about Mansa Musa, was more interested in Mali's past than in its present. In his spare time he picked through archives. One clue led to another until, in a library in Spain, he came across a 300-year-old map of the Sahara, and there on the map, in what is now Mali, Nathanson saw a name that would lift the heart of any treasure hunter: Ophir!

Treasure hunters have searched for a fabulous city called Ophir for thousands of years, hoping to find its legendary mines. In antiquity they looked for it in India, Arabia, and Africa. The son of a king of Sheba was said to have "built Ophir with stones of gold, for the stones of its mountains are pure gold." To the pre-Islamic Copts of Egypt, Ophir was another name for India, a country synonymous with opulence. Ophir is the fictional lost city of Haggard's *King Solomon's Mines,* and the name of a kingdom in the *Conan the Barbarian* series. In a famous lithograph from the California gold rush, a sailing ship arriving in San Francisco has the name *Ophir* on its stern.

How many quests start this way? The hero finds a map and off he goes. There was no Ophir on contemporary maps. Nathanson traveled through western Mali searching for towns that a cartographer 300 years ago might have labeled with the name of the legendary city. In that part of Mali artisanal miners were still producing small amounts of gold. Itinerant gold buyers regularly visited the area. Nathanson decided that if he were stopped by the authorities and questioned about his travels, he would say that he was scouting for things to buy, including gold.

Nathanson based himself in western Mali's provincial city, Kayes, a sweltering town of 100,000 surrounded by iron hills and baked at temperatures that often rise above 100°F. From Kayes he headed into the countryside along the Falémé River, the border with Senegal. On trips that lasted months, he visited the scattered villages. He inspected gold digs, hoping to find the remnant of a mine that might have once been rich enough to attract the name Ophir. In village after village he saw miners grubbing for small amounts of gold, with no sign that they had ever produced more.

With no particular expectation, Nathanson came to a village called Sadiola, an unpromising collection of farmers' huts. Scrawny cattle competed for grazing with scrawny goats. There was no other sign of village wealth. Nathanson explored the vicinity anyway. There was a hill nearby, and he set out to climb it. As he went up the slope he noticed an indentation. He stopped to investigate, and realized that the depression was the mouth of an adit, a horizontal mine tunnel. It was blocked with debris. No other signs of mining disturbed the hill, only the single adit closed by rubble. He returned to Sadiola with his guide, and learned the story of the hill.

There had indeed been gold mining at Sadiola. The villagers' ancestors had mined it for centuries. But about a hundred years before, the adit had collapsed, killing every man inside and decimating the local population. Mining had ceased from that day, the hill declared forbidden. No local person would dig there. The adit was what Nathanson had been looking for: evidence of large-scale mining, possibly important enough to have suggested the name Ophir to a European who had heard of it. Moreover, the tragedy and subsequent forbidden status of the ground explained why no one worked it now. To a gold seeker, Sadiola cried out for a drill.

But Nathanson kept quiet. With Soviet influence still strong in Mali, he knew that any discovery would end up in Russian hands. He bided his time. When the Soviet Union crumbled, Mali's government went looking for investors. And so did Nathanson.

"The first I heard of it," said Larry Phillips, a Toronto lawyer who became part of what its members called the Mali syndicate, "was a phone call out of the blue in late 1989. It was Bill Pugliese, one of my clients, and he was calling from Switzerland. He was on a ski holiday, but he had this deal that he wanted to proceed with. He sketched it out to me. A gold prospect in Mali. He wanted to buy exploration rights in western Mali. The Mali government was asking for a $2.2 million letter of credit from his bank, and Bill wanted me to examine the letter.

"I sort of panicked," Phillips said. "I thought—I hope he hasn't written a check! Then I thought, where's Mali?"

Soon Phillips was battling his way through the challenging due diligence of a deal with a country that had no advanced business infrastructure. Legal drafts flew back and forth across the Atlantic. Mali is a French-speaking country, and Phillips needed the help of French law firms with African expertise. As he learned more about Mali and its past, he got hooked. He read the journals of the eighteenth-century Scottish explorer Mungo Park and French colonial accounts of the region. He learned about the scale of the early gold mines. In the end, he found the prospect irresistible. "So I joined the syndicate myself," he said, "and we went to Africa to take a look."

One of the most important business practices of a small exploration company is secrecy. Big companies prey on little ones. From its earliest tactics the Mali syndicate showed how thoroughly they understood the need for stealth. To disguise the location of their target, they took a 500-square-mile exploration license—a massive package that a junior company could never explore properly. If pressed about the size of the license they wanted, they would say that they had multiple targets. Protecting their target became even more important when the government launched a program to promote exploration.

When the Russians had left, Mali had taken advantage of European Union development money to conduct a large-scale geophysical survey of the country. They wanted to locate mineral formations that would attract foreign explorers. The survey had identified the Kédougou-Kéniéba Inlier, where Sadiola lay, as a gold-prospective zone. Exploration companies arrived to take a look. The syndicate's large land position concealed its discovery. "We had all that land," said Phillips, "but really, the main target was always Sadiola, and that's what we did not want to advertise."

At Sadiola they found the galleries of a mine that had been worked for 900 years. By 1992 the syndicate had outlined a reserve of 3 million ounces. The mine went into production in 1996, planning for 285,000 ounces a year and an eight-year life of mine. Instead they got 400,000 ounces a year for twice as long, and have since found another 5 million ounces that will extend the mine life by a decade.

The discovery at Sadiola ignited an exploration rush on the Mali side of the inlier. Randgold Resources, a Channel Islands–based gold miner, found the Morila and the Loulo deposits, which together contained 15 million ounces of gold. With the liberalization of Senegal's mining code in 2003, the search swarmed across the Falémé River into Senegal. Teranga's Australian owners discovered Sabodala, and Randgold found a deposit at Massawa. As the gold price rose, this search intensified, and a day after leaving Dakar, Pawlitschek and I drove into the bamboo forest and reached Dalato camp.

THE CAMP SAT ON A high bank of the Falémé River. A row of square, white-painted concrete sleeping cottages with thatched roofs stood along one side of a central plaza. Construction workers swarmed around the concrete shell of a new accommodation building. The camp teemed with geologists and exploration crews, and was so overcrowded that drillers were spilling out of the cantonment into tent camps in the bush.

As we parked, the stocky figure of Donald Walker, the chief geologist on site, exploded into view. Walker and Pawlitschek plunged into a discussion of the day's complexities. Teranga had a 328,000-acre land package that stretched for sixty miles along the Falémé River. It was a largely trackless waste of choppy hills and rock and scattered villages. Lions and leopards patrolled the bamboo thickets. Onto this landscape the explorers were superimposing the logic of a gold search.

In the camp's main office, charts covered the walls. Cabinets with narrow drawers held maps and diagrams and drill plots. The geologists swept some Styrofoam cups from a counter and Walker spread out a chart that Pawlitschek wanted to examine.

The exploration of a large gold prospect typically begins from the air. A plane tows a pod containing a magnetometer, a device that scans the upper levels of the crust and measures the relative abundance of magnetic materials, such as iron oxide and magnetite. Because different rock types contain different amounts of these minerals, a map of the magnetism shows geologists where different rock types meet. Since the hydrothermal flows that carry gold from the mantle to the surface exploit such weaknesses as these meeting places, the mapping shows geologists where to look.

Another type of airborne survey narrows the search. Radiometrics measures the radioactive emissions from the surface that result from the decay of isotopes. This information helps date the rock. Geologists already know from regional experience the age of the local rock most likely to hold gold. This new data further help target the drills.

Geologists had already identified 10 million ounces of gold in eastern Senegal. Teranga had drilled hundreds of thousands of feet of exploration holes, and sampled and trenched throughout its license. It had twenty-seven priority targets. Late that night we went to see the richest one—the Gora target.

We left camp and drove past the sleeping village and into the bamboo forest. The trees beside the road were covered in dust and in the headlights looked like a forest of white trees. A panicked squirrel shot across the road, trailing a plume of dust, like snow. After a while a distant light came into view, and we made our way toward it, following the twists and turns of the track until we reached the fantastic scene where the drill rig hissed and roared.

Covered in white rock powder, the Ghanaian drillers looked like ghosts. They moved with an alien deliberation, like astronauts on the moon, filling plastic bags with chips of rock that gushed from the drill. They wore bandannas to keep from breathing dust. The drill was white and the men were white and the ground was white, and all around was the thick black night. A Senegal bush baby—a small, nocturnal primate—sprang from the darkness through the light, a pair of astonished eyes.

The Gora target was a gigantic block of buried quartz that poked out at that single point—like the tip of an iceberg. Gold-bearing "pay veins" ran through the quartz. One was thirty feet thick. The quartz went down in to the surrounding rock at a steep angle. The drillers had tracked it for about a mile. They knew where the gold veins pinched and swelled, but not where they ended.

The deeper they drilled, the more they found. On the night I was there the drills had outlined 70,000 ounces. At that night's gold price, the deposit would be worth $100 million. A few months later they had more than doubled the estimated gold to 156,000 ounces.

Now it's 374,000 ounces—enough for a five-year mine, and, as I write, more than a half a billion dollars worth of gold. In the forest, a rival army was watching to see what Pawlitschek had found, so they could find it too.

IN THE MORNING I RETURNED to the forest on a mapping trip with a short, grizzled geologist named Michel Brisebois, a Quebecer who had started out in life as a lumberjack. He'd developed a taste for roaming the world, and decided that the profession of geology offered the best way to finance it. We rode up with a camp employee, a South African army veteran who droned on and on, like a radio that could not be turned off. The program that morning was Horrible Things That Different Kinds of Ammunition Can Do to Your Body. He dwelt long and lovingly on the holes made by certain bullets: tidy hole at entry; messy hole at exit. Then the program switched to snakes and scorpions. In that part of Senegal they have the emperor scorpion and the black-necked spitting cobra. "But the worst is the puff adder," he said cheerfully as Brisebois and I piled out at our starting point. "The toxin of the puff adder is a cytotoxin. It attacks your cells. People do not always die, but they are never the same again." He gave us a smile like a bandolier loaded with white bullets.

Brisebois shot him a sour look and struck off into the bamboo. "That guy is an asshole," he said when we were in the thicket, "but I'll tell you something. Watch an African in the bush. He looks at the ground, because that's where the danger will come from. And don't walk too close to me. If I surprise a snake, you are right there on top of him before he has a chance to get away, and it's you who will get bit."

Brisebois wore a tan vest that bristled with pens. A compass dangled from his neck. His graph-ruled notebook filled with tidy entries as we picked our way among GPS coordinates. The day was fresh and the forest suffused with a straw-colored light. A faint smell of wood smoke lay on the air from farmers clearing land. As we made our way through the greening thickets, charred branches striped my shirt with ash.

"The burnt area is very efficient because you can walk quickly through it and see the rock," Brisebois said when we reached a formation. "Other than these outcrops, we are walking on a thick cap of laterite. Beneath that lies the gold host. This ridge marks a shear zone. The sedimentary rock sheared and these quartz ridges popped up through it, giving evidence that more of them must lie below."

Brisebois loved to handle rock. He picked up a piece and opened it with the barest tap of his pick. "You see these boxlike shapes?" The pale gray rock was speckled with faint, rust-red outlines winking with tiny flakes. "They are an iron sulfide called pyrite—the fool's gold that many people recognize. In this deposit, the real gold is associated with the fool's gold."

On the ridge, the bamboo broke the sunlight into splinters. A drill roared nearby. A backhoe had made a trench in the hill we were exploring, and there we chanced on an outpost of the competition. At the bottom of the trench gaped a deeper hole, about a yard square; but this one had been dug by hand. A well-made buttressing of logs kept it from collapse. The shaft penetrated too far down for me to see the bottom. The local people who had sunk the hole had viewed the backhoe trench as a free head start. They had reasoned that if the geologists thought there might be something there, it was worth a look. Sometimes the reverse happens—explorers sample where the local people have been digging. A cat-and-mouse game had developed in the forest as each of the two bodies of experienced gold searchers, the new and the old, circled each other. That afternoon I drove out of camp with Djibril Sow and Thierno Mamadou Mouctar, Senegalese geologists on Teranga's staff, to see if we could find where villagers were digging.

We followed a red road that rose and fell through the hilly forest until it climbed to a plateau and brought us to the village of Bondala. Most of the houses were small and round, covered in a stucco of mud and topped by conical roofs of trimmed thatch. Decorative patterns of twine, ornate and beautiful, held the thatch in place. Bamboo fences encircled the village and divided it into compounds. The only exceptions to the mud construction were a pair of concrete structures—the village school, and the house of a villager who had found a one-pound lump of gold. Bondala was a mining town.

A rough track from the village led into a bamboo thicket. The truck plunged and bucked through crater-sized holes, throwing up clouds of powdery dust that billowed across the hood and covered the windshield. The driver put the wipers on to clear the dirt. We rolled up the windows, but soon our clothes were coated with a fine dust. The geologists said it was a kind of silt, evidence of the sedimentary soils that host the gold in some parts of the prospecting area.

Termite mounds ten feet high rose in the thin wood. Baked hard by the sun, the mounds are made of excavated earth carried up from the termites' tunnels. The deep systems can extend to sixty feet below the surface. Geologists regularly sample termite mounds to see what lies below. Teranga's field workers had sampled 20,000 mounds. One of the richest mineral discoveries in history, Botswana's Orapa diamond pipe, was located with the help of mineral clues carried up by termites. Sampling insect hills may be an ancient practice. In the fifth century BC, Herodotus recorded a story about ants "bigger than a fox" that excavated gold-rich sand in the deserts of Afghanistan.

We came to a place where Teranga crews, investigating reports of local gold digs, had come in and trenched with a backhoe. Gravel lay in heaps, and the tattered forest showed where tracked vehicles had blundered through. Finding nothing promising, the crews had left. But trenching prompted the villagers to investigate in turn. They had put up bamboo sunshades to shield the trench, and dug a few holes. They had not found anything worth pursuing either, and all that remained of their activity was a broken calabash and some poles.

There was not a soul in sight. Djibril Sow was certain there was mining somewhere near. He pointed to fresh tire tracks that led down a narrow track. "Their water truck," he said.

A cyclist came wobbling along from the direction of the village. We called out but he pedaled away into the bush, ignoring us. We left our truck and followed him on foot. Soon we saw the water truck, more bicycles, and came to the dig. At first I saw only a few dozen people working around six or seven holes, but as we walked into the site, I saw that it stretched much further, as far as I could see into the thin wood. The whole village was out digging for gold.

Clearly they resented and distrusted me, and would not even look at me until my companions explained that I didn't work for the company and was not a geologist. It would have been hard to mistake me for one. I wore an old tennis hat. My hands were as red as lobster claws and my snowy outfit, fresh from the hotel laundry in Dakar, was striped with ash from top to bottom. The kids thought I was hilarious. They would dash up and shout at me from a few feet away, and when I looked at them, they would whirl away and race off squealing. I jumped at one, and the whole pack went shrieking into the trees; but I was fair game after that.

I watched one man scuttle down a hole at least twenty feet deep. He had no ladder or rope, but went down by bracing himself against the shaft wall. With a few quick movements he was at the bottom, where he disappeared. A network of deep tunnels ran beneath the wood. A steady feed of ore came up in baskets. The women panned it in their shallow calabashes. They manipulated these with mesmerizing skill, swirling the water and soil around and tipping off the lighter particles until they had a residue of heavy grains, and sometimes, sparkling among them, a globule of gold. "They call it in their tongue *nara*," said Djibril Sow. "It means nugget."

GOLD IS ITS OWN COUNTRY. One morning on the bank of the Falémé River I watched a motorcycle buzz into view from the bush on the Mali side. The rider navigated down the muddy slope, bounced across the shallows and tore off into Senegal, headed for a sprawling artisanal minesite at a place called Soreto. It dwarfed the forest digs at Bondala, and miners from Mali commuted to it every day.

The mine was a fairground of men and women, laborers and vendors, miners, children, dogs. The women and girls blazed with gold earrings. People greeted us with shouts. *"Bonjour! Ça va?"* A woman sipping from a glass of yellow liquid raised it to us in a toast and scorched us with her smile.

They were mining a strike that ran for half a mile. They had trenched the length and screened it with bamboo shades. From the floor of the trench, shafts went down to the mining galleries. Some of the shafts reached depths of 100 feet. Sometimes they hit water, and the miners clubbed together to rent pumps. On a rough head count, about 300 people were working in the mine. Many more supported it.

Mechanics serviced the pumps and blacksmiths made tools. The smiths pumped their bellows with one foot while hammering at iron implements on their anvils. Vendors sold popsicles and water, cigarettes and candy. A small solar array powered a battery-charging station for cell phones. Masses of bicycles and Chinese motorcycles leaned together under a thorn tree. A man with an air pump between his knees sat in the scanty shade repairing tires.

The whole site, and others like it, belonged to the village of Diabougou on the Falémé River. The chief levied "license fees" on miners who were not native to the village, which was most of them. The hamlet had swelled from a population of 1,000 to about 10,000—a rapid influx of outsiders drawn by the gold boom. Diabougou was a boomtown. In stalls along the widest thoroughfare you could buy shoes and shirts, blankets and mattresses, plastic toys, vegetables, fish, televisions. Two cell phone dealers competed head-to-head across the street. On one side of the town stood the old village of round mud houses with thatched roofs; on the other, dwellings bashed together out of anything at hand: mud, tin, planks, cardboard, vinyl sheets. Scooters buzzed through a labyrinth of lanes and boys toiled up from the river with handcarts loaded with plastic water jugs.

Inexorably, the gold rush was erasing an old way of life. On the way to the town we'd met a large herd of goats filling the road. The goatherd was a young man in outlandish costume. He wore a round black hat with a narrow brim, and a loose, ragged skirt almost to his ankles. Tall and thin, he looked at us with profound astonishment. He emitted a series of short, low whistles, like a sentence of code, and the goats surged off the road. He stopped in the grass and gaped at us in bafflement as we went by. I was told these herders have roamed immemorially through Mali and Senegal, following the grass, and are now disappearing. They mine gold instead.

On the flats beside the river stood a line of sluices fed by water pumps. To recover gold, they directed water down a sluice and shoveled ore in at the top. The water carried the ore over strips of carpet nailed to the bottom of the channel. Light soils flowed away while the heavier, gold-bearing gravels snagged in the carpet. They would use mercury to concentrate the gold, handling the toxic substance with bare hands. According to Moussa Bathily, a Teranga geologist, Friday was the day reserved for this gold recovery. No one would work at the minesite on Friday. I asked him if that had anything to do with Friday being the Muslim holy day. "Oh, no," he said. "They leave the mine because they say that Friday is the day the devil comes to put back the gold."

12

KIBALI

On one side of the contest were murderous, well-armed brigands in possession of the world's original cash crop—gold. On the other side: a powerful mining company.

ON A FEBRUARY DAY IN LONDON I SET OUT TO WRITE this book. I flew to Entebbe, Uganda, where I spent the night at a hotel on the shore of Lake Victoria, battling a cloud of gnats that were feeding on my ears. I had Peter Bernstein's book, and riffling through the first pages I spotted a reference to the queen of Sheba bringing gold to Solomon. I began to read, but the text dissolved in a veil of pests, and I gave up and covered my head and tried to sleep.

I was on my way to a discovery much richer than the fabled queen's mines. In the gold world everyone was talking about Kibali, a breathtaking target in one of the most benighted countries in the world—the Democratic Republic of Congo. Like Congo itself, the deposit had a tortured past and a future made tantalizing by the abundance of its riches. In its century of history the Kibali goldfield told a story of oppression and war, of old money and vested interests, of businesses stalking each other. Dazzling technologies went side by side with destitution. I end with it because it captures the thrill of a modern gold rush, its speed and ruthlessness, and also because it shows how unsettling and barbarous gold can be, and then again, how fortunate to find. In Kibali were all the sadness and excitement of the human condition.

In the morning I went back to the Entebbe airport and squeezed into a single-engine Cessna Caravan stuffed with overheated mining stock analysts. They had come straight from a grueling tour of gold mines in Ivory Coast and Mali, and except for a cheerful, fresh-looking young woman from California, they were all damp and grumpy. We took off for Eldorado.

The Kibali gold project lies in the northeastern Congo, a war-torn region only partly pacified by United Nations peacekeepers. The Lord's Resistance Army, a killing machine of drug-addled child soldiers based in Uganda, was still kidnapping and murdering in the province when we visited in early 2011. We cleared Congolese formalities at Bunya, an airport bristling with the machine gun nests of the U.N. force. An elderly man at a table in the airport building received my $90 cash payment, and slid my passport to another functionary, who stamped me into the Democratic Republic. From Bunya to Kibali was a twenty-minute flight.

The target lay in a dejected valley scarred by exploration tracks. From a hilltop vantage point beside the cell phone tower we could see the crumbling huts of a village on the far slope. A drill rig rattled on the hillside. The equatorial sun beat down through a layer of thin gray cloud. The mud of a drained lake bed had puckered and hardened in the heat. Beneath the wrinkled surface lay 10 million ounces of gold.

• ● •

AUSTRALIAN PROSPECTORS HAD DISCOVERED THE gold in 1903, on a hundred-mile-long formation called the Kilo-Moto greenstone belt. Congo was then the personal possession of the Belgian king. Until Congolese independence in 1960, the mines produced about 12 million ounces. After independence, the new government transferred the mines to a state-owned company. It struggled to operate the mines. It looked for foreign partners. Barrick operated a mine near the present discovery, until the war arrived.

In 1996 Rwandan and Ugandan troops invaded Congo to dislodge the tyrant Mobutu Sese Seko. When Mobutu fled, his successor tried to expel the foreign armies, which were feasting on Congo's rich resources. A bloodbath followed, drawing in forces from Zimbabwe, Burundi, Angola, and Namibia. In a rampage often called the First World War of Africa, 35,000 foreign soldiers murdered, raped, and pillaged their way through the country. They stole diamonds, oil, uranium, even parrots. They slaughtered rhinoceroses for their horn. The spoils of war poured out of the Congo. In 1998 the conflict reached Kibali.

In August 1998 Ugandan soldiers occupied the goldfield. They had planned to run the mines as industrial concerns, but found the task beyond them. They resorted to artisanal mining—basically, smashing the rock to bits with hammers and picks, and sometimes explosives, and carting out the ore by hand. They drafted local miners to the work, charging them fees to enter the site or collecting a cut of the ore. They beat miners who refused to work. They forced them to blast out the gold-rich pillars of rock that had been left in place to stabilize the mine. As a result the mine collapsed, killing a hundred miners trapped inside. The collapse knocked out the pumps. Water flooded the mine and valley floor, forming the lake whose dried-out bed still marked the site when I was there.

Things got worse in 2002, when international pressure forced Uganda to begin withdrawing its forces from Congo. As they did, rebel militias moved onto the goldfield, bringing to the brutalities already in place a campaign of terror driven by tribal hatreds. In one massacre, women and children dug their own graves. Then their captors roped them together and butchered them with machetes and sledgehammers. The conditions in the mines were horrendous. Miners died of suffocation in tunnels once ventilated by air circulation systems that had fallen into disrepair. Similarly, failed pumps meant that miners had to wade for miles through chest-deep water in narrow tunnels to reach the mining galleries. Once there, the men used iron bolts and ordinary hammers to batter ore from the stopes, and lit fires to "soften" the rock.

The mines delivered rich returns to the usurpers. The Ugandans took $9 million worth of gold from the district in four years of occupation. Their successors did even better, with one armed group selling as much as sixty kilograms of gold a month, worth three quarters of a million dollars. For circumstances had changed. After three years of slumber the gold price had awakened. The Ugandans had been getting $290 an ounce from their Swiss buyers; the militias got more than $400. The gold price had just set out on what would become the most phenomenal streak in its history, the ten-year bull run that set off the world gold rush. This was a two-edged sword for the plunderers. In the short term it put more money in their pockets, but in the long term it attracted the attention of forces that would want the gold themselves. One of these was a black African tycoon.

· · ·

SIR SAM JONAH WAS A celebrated gold miner who had started out in his teens as a mine laborer in Ghana, studied mine management in Britain, and by the age of thirty-one had become the chief executive of Ashanti Goldfields, a Ghanaian miner that he ran for eighteen years. In a single decade he raised Ashanti's production from 240,000 ounces a year to 1.6 million ounces, turning it into a top gold miner, and, in 1996, managing to list it on the New York Stock Exchange. He had his setbacks too. In 1999 a wrong-way bet on the gold price almost ruined the company. But Jonah pulled Ashanti through. The Prince of Wales knighted him in 2003. In 2004 the magnate merged his company with AngloGold Limited, a South African miner owned by the Anglo American group. The merger created what was at the time the world's second biggest gold miner, AngloGold Ashanti.

The next year, 2005, Jonah resigned his executive position at the new company, although he stayed on the board. Among the reasons he gave for leaving was the wish to pursue other ventures. One of those was a company called Moto Goldmines Limited, a Toronto-listed junior looking for gold at Kilo-Moto. They were going to find a very large amount of it, as Jonah was in a good position to guess. He had taken Ashanti into Kilo-Moto nine years earlier, in 1996, when the company bought into a joint venture with the Congo state-owned company OKIMO to explore and mine a 3,800-square-mile concession.

Let's pull back to a bird's-eye view and consider the field. Below us is a geological formation long considered one of the world's best bets for gold. Explorers knew its potential, but hadn't staked it out and scrutinized it because the gold price wasn't high enough to compensate them for the risks inherent in making a discovery, namely: the runaway corruption of Mobutu's regime, or later, the fact that northeastern Congo was a terror-stricken killing ground. But in

2003 a national government was forming in Kinshasa with international support, giving hope that Congo's war would end. Mining companies began to maneuver for a place on the Kilo-Moto.

I suppose it all got a bit lurid, what with mining companies drooling at the riches, and the homicidal gangsters actually running the digs showing up for meetings in Kinshasa. AngloGold funded one such trip, and Human Rights Watch accused them of giving the killers "material benefits and prestige." The company said it had only paid $8,000 out of petty cash, and had done it mainly to protect its own employees. Anyway, the parties worked things out. The killers got commissions in the Congo army and Moto Goldmines got Kibali.

I have said that Moto was a "Toronto-listed junior," which might sound inconsequential. But Toronto is the world's leading gold-mine city. Some sixteen hundred miners list their shares there. It has the world's largest pool of capital for mine finance. More than $280 billion worth of mining stocks traded on the Toronto Stock Exchange in 2012. Almost three quarters of the equity raised in the world for new mining ventures that year came from Toronto. More than 500 analysts in the city follow mining stocks, and three of the world's biggest gold miners are based there. "Junior" means a small mining company focused on exploration, as opposed to a senior, like Barrick, that operates large mines. Regardless of its size, Moto was the cat's paw of important interests—Sam Jonah, AngloGold, and through arrangements with OKIMO, the Damseaux family, Belgian industrialists with a cattle ranching and transportation empire in Congo dating back to 1931. Moto was the mailed fist of power, never mind the glove.

They drained the lake and drilled the bed. They hit gold, and kept on hitting it. Their press releases sang out on the business wires, confirming the promise of the Kilo-Moto greenstone belt. The gold price rose. Moto's value rose. One industry website called it "the hottest gold stock in the world." Now it was time for the next part of the drama. Moto had gobbled up the goldfield, and now they would get gobbled up themselves. The fight for Moto pitted two gold-rich tycoons against each other in a battle that the victor quarterbacked from the saddle of a speeding motorcycle.

"WE BEGAN TO WATCH MOTO in 2006," said Rod Quick, the chief of exploration for Randgold Resources, a Channel Islands–based gold miner. "I tracked their data hole by hole. I plotted it. I visited the site."

Both companies understood this relationship perfectly. A junior company exists to attract such predators. The prey hopes—intends!—to be caught and swallowed whole. It's important for the hunter to remember this. "Each time we got a drill result from them I'd put it into the mix," said Quick. "Obviously, juniors are only going to report their best holes, so you have to factor in what you know are the lower grades and make an estimate based on that." Even so, the judgment was that Moto had outlined a deposit of at least 5 million ounces of gold.

"The first thing we did," Mark Bristow, Randgold's chief executive, told me at a meeting in Cape Town, "was to get the [Congolese] government to buy in. I told them, 'If you are not supportive of our bid, we will not go ahead. We promise we'll build a mine: here are our plans. We are launching a hostile takeover [of Moto]. Are you behind us? We don't want a fight. We don't want an auction. We want to kill it.' "

Bristow is a fifty-two-year-old South African with a Ph.D. in geology and the build and temperament of a Cape buffalo. He has a weakness for such diversions as hurling himself out of airplanes for a thirty-second free fall; shooting Grade-V rapids on the Zambezi River; and bungee jumping at Victoria Falls, where you drop ninety-five feet before you reach the end of the cord. He has homes in London, Johannesburg, Mauritius, and the ski resort of Jackson Hole, Wyoming. But mostly he lives in a succession of airplanes, crisscrossing Africa on the hunt for gold.

With its joint venture partner, AngloGold Ashanti, Randgold took a run at Moto in February 2009. With $258 million in the bank, the South Africans were "cashed up." But they had competition—Lukas Lundin, a Vancouver-based Swedish minerals magnate. One of Lundin's companies was Red Back Mining, which at the time was riding high on the success of its Tasiast gold discovery in Mauritania, a property swelling in value as the gold price rose through $1,000 and the size of the deposit grew by bounds. Red Back made an all-paper bid—an offer in which the currency was Red Back stock. The offer valued Moto at $486 million. "There is no question in our minds that [it] is a world-class gold project," Red Back said.

Randgold responded by taking a "blocking stake" in Moto—a technical maneuver allowing it to block the Moto board from accepting the Red Back offer. Moto was listed in Toronto, in the province of Ontario. Under Ontario securities law, a takeover needed the support of two thirds of shareholders. Randgold had managed to get voting control of just over a third of Moto's stock, effectively preventing the two-thirds quorum from forming. Randgold did not own the shares, but could vote them through a "soft lock" agreement with the actual shareholders. A soft lock commits the participating shareholder to support a course of action, but only to a point. That point, for example, might be an even better deal than the one Rand-gold was contemplating. In such a case the lock would dissolve and the shareholder would be free to take the better offer. Randgold's control was thus provisional, but it prevented the Moto board from accepting Red Back's offer right away.

Bristow worked on his counterbid while roaring north up Africa on a forty-nine-day motorcycle trip from Cape Town to Cairo with his two sons and three of his friends. "We had a huge intercom system on all the bikes," said Grant Bristow. "Helmets were wired so we could take cell phone calls individually by satellite. But if you used the intercom to talk to somebody else, you overrode any incoming phone call. So there would be these extended periods when Dad was doing business and no one was allowed to talk."

They were riding heavy BMW bikes with special after-market shocks to handle the brutal roads. "In northern Kenya there was this massive corrugation from the truck convoys, and four of the bikes blew their shocks," Grant Bristow said. Worse than the discomfort was the danger: Sudanese bandits lie in ambush for the convoys on that stretch of road.

The riders made it through. Bristow cobbled together a half-stock–half-cash offer that was $30 million sweeter than Red Back's. By the time the riders got to Addis Ababa, Randgold had Kibali in its pocket.

Cashed-up tycoons were the order of the day. Like Bristow, they all seemed to glow with the oxygen of extreme sports. Red Back's Lukas Lundin, the scion of a Swedish mining-and-oil dynasty whose founding asset, Lundin Petroleum AB, had rivaled Apple Inc.'s spectacular stock performance by multiplying its value fifty times since it had started trading in 2001, had climbed Mount Kilimanjaro, competed in the Paris-Dakar motorcycle race, and liked to plunge down thickly wooded mountainsides in pursuit of the thrill of extreme skiing. When I was in Cape Town at a mining convention after the Kibali visit, the halls were thronged with people debating whose was the mine of the century—Bristow's Kibali or Lundin's Tasiast, although by then Lundin's role at Tasiast was that of shareholder in a company he did not control. He had sold the Mauritanian property for a pack of stock in a company run by another high-octane mogul.

Tye Burt, who bought Lundin's gold mine, was a 53-year-old yachtsman, lawyer, and investment banker. He ran Kinross Gold Corp., a Toronto miner that vaulted into fifth place among world gold producers by paying $7.1 billion for what he believed to be the eye-bugging reserves of Lundin's mine. Burt projected an air of unshakeable self-possession. He had a taste for the fiction of Ernest Hemingway. In Paris, he liked to drop in at the Café de Flore, a Hemingway haunt. Taking things up a notch, he had also run with the bulls three times at the festival of San Fermin in Pamplona, Spain, a rite immortalized by Hemingway in *The Sun Also Rises*.

"The bulls run three kilometers," Burt told me, "and the side streets are closed to traffic. So the route becomes a sort of chute. The streets are packed, and they throw up the barricades and fire a cannon, and the bulls come out. You can't outrun them—they are going full clip. So what you do is try to dodge them. If you really want to be cool, you swat a passing bull with a rolled up newspaper."

Anyway, there you go: ran with bulls. The market punished Kinross stock, judging that Burt had overpaid for an upside that might not be there. "I hear they have 20 million ounces," I put to Mark Bristow. "Sure," he said, "and tomorrow—30 million." Kinross fired Burt in August 2012, after the company took a $3.1 billion writedown on the Mauritanian property, two years after buying it.

IN CONGO, THE HIGH GOLD price spreads its benefits unevenly. At Kibali a 110-mile highway through the bush now links the province to the outside world. In come cheaper goods. Randgold hires local people, nurtures small businesses, has built a new town, church, school, clinic. Chinese motorcycles multiply like loaves and fishes, and gold bars will clunk from the plant for the investment payoff. A United Nations force helps pacify the region. Its soldiers come from countries paid to field them. In a sense they are security guards hired to protect property rights. If this arrangement seems cynical, consider the alternative.

A killing machine grinds its way through Congo, seeking gold. Tens of thousands of artisanal miners suffer a relentless toll. Soldiers, often unpaid by the government, demand cuts of gold production. Sometimes the soldiers use gangs of young men to terrorize the digs. A woman with a small store near a gold dig told researchers that it made no difference to the people whether it was a militia or government troops in charge. "They pillage, they rape, they kill, and they force us to give them money all the time. We have no peace, no matter who controls the region."

To escape extortion by officials, miners sell their gold into the illegal trade, a much larger commerce than the legal one. As much as $400 million worth of contraband gold leaves Congo every year. In one six-month period in 2012, when official figures for the eastern Congo listed exports of twenty-three kilograms of gold, the true figure was as high as four metric tons.

The booming illegal trade made gold the main source of income for the killing posses of the eastern Congo. Because the United States' Dodd-Frank Act makes companies account for the origin of minerals they buy that might come from the war zones of central Africa, militias find it harder to sell such minerals as tin or tungsten, but easy to sell gold. A human rights group calculated that $30,000 worth of gold would fit in a pocket and $700,000 in a briefcase. In eastern Congo, a war that has already killed 3 million people threatens to break out again, fueled by ethnic hatred and by gold.

One afternoon I sat with a modeler at Kibali and gazed at a vivid 3-D computer image of the main ore body. Beneath the undulating valley floor sprawled the gorgeous, tentacular, magenta-colored shape, with everything drawn in—the ventilation and mining shafts and the spiral tunnels for the trucks. A Randgold tech revolved the image on the screen to display it from every angle. It was hatched with thin lines that showed where the geologists had drilled, hole after hole, tracking every twist and turn of the ore.

Mining companies work in an environment swept by gold fever without catching the virus. They know that price is an unstable factor. They protect themselves by pricing their reserves well below the spot price. The Kibali mine will cost Randgold and its partner $2 billion, an investment that could be destroyed by a turn in the gold price. Randgold priced its reserves at $1,000 an ounce, meaning that they don't consider rock to be ore unless it can be profitably mined at that price. Because price is so crucial, miners think about it all the time. I raised the subject with a South African gold executive, who asked me not to use his name because he was connected to the Kibali project and did not want his speculations on the record.

"With consumables, it's easy to judge price," he said. "If one pair of shoes costs $200 and another pair $300, you compare them and make your choice. If you want a bond, you compare interest rates. But if you decide to buy gold you pay the spot price. There is no reference, no other thing like it to compare it to." He had studied at the Wharton School and had managed funds at a Swiss bank. "My whole life as a businessman I have struggled with questions of value," he said. "It's easy for me to engage in negotiations about value when there's a reference point. But what's the reference point with gold?"

The only reference point is other assets. Fear that other assets would not hold their value started gold's phenomenal price rise in the first place. That rise accelerated with the banking crisis that began in 2008, but it was already under way. From 2004 to 2007, investors who worried about the sustainability of the boom in such asset prices as stocks and real estate, bought gold to hedge against a collapse. Gold investment in that period—around 600 tons a year—was twice what it had been in the four years that preceded it. The question of how far the gold price can rise becomes a question about how far down the economy can go. Or put the other way—the gold price will flame out when the economy stabilizes.

Except for a few practical uses, gold is a notional construct. It has no meaning but its price. Even the jewelry market hangs on that consideration—whether gold as a material is worth more than something else. Gold has moved far from its original place in the human imagination. It's not clear that our distant ancestors even thought of it as valuable in the material sense. For much of prehistory, gold was placed in the ground as votive offerings. Our ancestors put gold in rivers from 2500 BC to 800 BC. Obviously they valued it, because it had some ceremonial role, but its value was not necessarily for trade. It was a value, scholars think, not for this world, but for the next. Gold was the last good-bye, a wish that would have to last forever. Maybe its immutability stood for the endurance of the human spirit in the face of death. Something of that association must have come down to us, and in the face of another apocalypse, people reached for gold.

In the financial crisis of the twenty-first century, doom was in the air. One theory about the super-rich saw in the growing concentration of wealth the ultimate destruction of the class acquiring it. As they drained more and more of the available resources into their own pockets, impoverishing the other participants in the economy, they were killing the economy itself, and hence themselves. In the steady procession of awful stories through the news, there was a sense that the perpetrators of the disaster had not changed their ways. Banks engaged in criminal activities, including money laundering, interest-rate rigging, and illegal home foreclosures. As a result of the limping economy, government revenues decreased and public finances deteriorated. In this environment, gold had its best bull run in history.

Every week a file of JPEGs lands in my inbox detailing the progress of the 45-million-ounce gold mine at Oyu Tolgoi in Mongolia. In the bamboo forest, thousands of diggers burrow through the soil. Now the search has moved to the deep ocean, where trillions of dollars worth of gold lies in sulfide deposits at volcanic vents. Nautilus Minerals of Toronto has a twenty-year lease on a target in the southwestern Pacific that it estimates to hold ten tons of gold. The Chinese are developing a submersible to explore deep-sea gold deposits and have acquired the rights to 3,860 square miles of seabed on a two-mile-deep volcanic rift in the Indian Ocean. There is no new use for gold driving this search, just the old one: an inextinguishable conviction that it will always save the day.

Like other epidemics, gold fever sweeps the world with unequal effect. At Kibali, we drove one day through Durba town in a cavalcade of white SUVs, rolling clouds of thick red dust onto the people by the road. Our windows were rolled up tight against the heat and dust. From our air-conditioned spacecraft, we looked out at the alien souls. Many wore rags. Ninety percent of the children had malaria. Some villagers would escape this in a miraculous transformation, moving into the houses that Randgold was building and that we were on our way to see. There were ocher bungalows with glass windows and doors that locked, and water and electricity, laid out in a new town in a palm grove by the sparkling river. The position of a fence decided who would live there. The fence enclosed the gold mine property. Land that Randgold wanted to dig up was inside the fence. If your house was inside too, you moved to paradise.

Outside the town I saw a woman standing in a field, leaning on a rake. She wore a green turban and a dirty green dress, and from her face, every shred of hope had been extinguished. She turned her back as we went by.

Acknowledgments

Special thanks to Barry Eichengreen for reading the manuscript and making valuable observations. Thanks also to Michael Woodford of Columbia University for explaining the mechanism of the gold standard, and for reading parts of my account. Neither of these scholars is responsible for any blunders I may have made.

I owe a large debt to Kathy Sipos at Teranga Gold for arranging my trip across Senegal, and to Martin Pawlitschek for his time and patience. At the Teranga exploration camp I was lucky to be shown around by Donald Walker, Djibril Sow, and Thierno Mamadou Mouctar. Thanks to Mark English for a great visit to the Sabodala mine. I would be lower than a churl if I failed to thank Awa Ba for giving me a lift out of the pit in her 100-ton ore truck. Of course I owe most for the Senegal visit to Alan Hill, Teranga's chief executive, not only for the visit, but also for his generosity in reading parts of the manuscript and correcting technical mistakes; for sharing with me his lively recollections of some of the most exciting passages in modern gold mining; and for giving me lobster for breakfast on a terrace overlooking Table Bay in Cape Town, allowing me a glimpse of what life is like for those who run gold mines.

Warm thanks to Kathy du Plessis at Randgold Resources for finding me room on a small and crowded plane, and to Rod Quick and Paul Harbridge for taking pains to explain the complexities of the ore body at Kibali.

Greg Hall opened many doors for me in China. His affection for the country, and for the distinguished gold people who are his friends, helped me understand their remarkable feat. Special thanks to Professor Zhu, to Feng Tao, and most of all to X. D. Jiang, that indefatigable practitioner, who wrung a profitable modern gold mine out of the most unpromising material.

Hayden Atkins at Macquarie Bank in London and James Steel at HSBC in New York helped me understand the bullion analyst's perspective, and I often talked to Sterling Smith of Country Hedging. I am very grateful to my cousin, the investment banker Mark Cullen, for finding someone to verify the mechanics of how a hedge fund might manipulate the gold price, and for his generosity in introducing me to market insiders whose identity I have agreed to protect.

My introduction to Bad Brad Wood came from Sally Evans, a star reporter for the M&G Centre for Investigative Journalism in Johannesburg. Stefaans Brümmer, a veteran reporter and a managing partner of the Centre, put me in touch with Sally. The Centre is funded by, among others, the Open Society Foundation and the *Mail & Guardian* newspaper, which carries the Centre's reports.

Alan Fine arranged my visit to Mponeng, for which many thanks, with special gratitude to Clive van der Westhuizen, the mine's engineering manager, for diligently answering my innumerable follow-up questions.

ACKNOWLEDGMENTS

I owe much to Dean Heitt at Newmont for showing me around the original mines of the Carlin Trend, and for reading my chapter on the discovery of invisible gold and offering suggestions. Anything amiss in the account is my fault entirely.

I thank Andy Lloyd for setting up my first interview with Peter Munk and arranging the visit to Goldstrike.

Thanks to my dear friend Alex Beam and to my old comrade-in-arms, Ian McLeod—thanks for the push.

Most important, for withstanding a withering fusillade of drafts, and for keeping up a steady, level-headed, and unnervingly accurate return fire, my deepest thanks are to my wife, Heather Abbott.

NOTES

CHAPTER 1: THE UNDERGROUND METROPOLIS

PAGE

1 *Picture Manhattan Island:* Google Maps gives distance from 59th Street to 110th Street as 2.7 miles. AngloGold Ashanti's reserves profile ("Ore and Reserves," pdf at http://www.anglogold .com) for 2011, p. 35, confirms mining to 126 level (12,600 feet = 2.38 miles). Exploration drilling extends lower. In 2013 they are drilling from 126 level to hit the deeper Carbon Leader Reef at 4,200 meters (2.6 miles). The headframe completes vertical silhouette of mine to ±2.7 miles. All other physical mine data from reporting.

2 *Their target was a thirty-inch-wide strip:* See http://www.mining -technology.com/projects/mponeng: average channel width 78cm, or 30.7 inches. For gold prices see http://www.kitco.com/gold .londonfix.html. For value of deposit: annual production of 600,000 ounces reported at http://www.infomine.com/minesite/ minesite.asp?site=mponeng, multiplied by that day's London morning fix of $1,581.

2 *The world is awash:* 2011 survey of bullion market: Jack Farchy, "Sizing Up the Gold Market," *Financial Times*, September 9, 2011, http://www.ft.com/intl/cms/s/0/eb342ad4-daba-11e0-a58b-00144feabdc0.html#axzz2QoQC31wu.

2 *As the gold price soared:* "Soros Doubles Down on Gold," *New York Times*, February 2, 2010, http://dealbook.nytimes.com/2010/02/17/soros-doubles-down-on-gold/; Azam Ahmed and Julie Creswell, "Bet on Gold Nets Paulson $5 Billion," *New York Times*, January 29, 2011, http://www.nytimes.com/2011/01/29/business/29paulson.html.

2 *Fear drove the price:* Allan H. Meltzer, "Gold Fever Is a Symptom," *New York Times*, August 2, 2011, http://www.nytimes.com/roomfordebate/2011/08/02/should-central-banks-buy-gold/gold-fever-is-a-symptom-of-inflation-fears.

3 *Sometimes the quakes:* See for example Dennis Ndaba, "Can South Africa Stop the Mine Fatalities?" *Mining Weekly*, February 1, 2008, http://www.miningweekly.com/article/can-south-africa-stop-the-mine-fatalities-2008-02-01.

3 *Some of the rockbursts had been so powerful:* John Oxley, *Down Where No Lion Walked* (Johannesburg: Southern Book Publishers, 1989), 159.

4 *Sometimes it winds men to their death:* Robert Block, "Locomotive Crushes 105 Gold Miners," *Independent*, May 12, 1995, http://www.independent.co.uk/news/world/locomotive-crushes-105-gold-miners-1619145.html. Liezl Hill, "Nine Killed in Accident at Gold Fields' South Deep mine," *Mining Weekly*, May 1, 2008, http://www.miningweekly.com/article/nine-killed-in-accident-at-gold-fields039-south-deep-mine-2008-05-01.

4 *Once our cage was full:* All data on winders from Clive van der Westhuizen, engineering manager, Mponeng mine.

8 *Swarming the gold mines:* My account of ghost miners is based on interviews with police and mine officials, and on-site visits, but see also "100s of Miners Could Be Buried," News24, June 4, 2009, http://www.news24.com/SouthAfrica/100s-of-miners-could-be-buried-20090604; Monako Dibetle, "Dying for Gold," *Mail & Guardian*, June 15, 2009, http://mg.co.za/article/2009-06-15-dying-for-gold; "Mystery of Aurora Corpses," *Mail & Guardian*, August 13, 2010, http://mg.co.za/article/2010-08-13-mystery-of-aurora-corpses.

8 *Gold once had a sacred aura:* For Charlemagne's reliquary, see http://www.sacred-destinations.com/germany/aachen-cathedral. For St. Edward's Crown, see http://www.royalcollection.org.uk/collection/31700/st-edwards-crown. For the Seville altarpiece, see Francisco Gil Delgado, *Sevilla Cathedral* (Barcelona: Editorial Escudo de Oro, 2003), 28–33.

9 *In August 2011 the "BlackBerry riots":* "The BlackBerry riots," *Economist*, August 13, 2011, http://www.economist.com/node/21525976; Josh Halliday, "London Riots: How BlackBerry Messenger Has Been Used to Plan Two Nights of Looting," *Guardian*, August 8, 2011, http://www.guardian.co.uk/media/2011/aug/08/london-riots-facebook-twitter-blackberry; Richard Partington and Jennifer Bollen, "Square Mile on Alert over London Riots," *Financial News*, August 9, 2011, http://www.efinancialnews.com/story/2011-08-09/bank-branches-left-closed-damaged-by-london-riots.

10 *His first client:* For Mandla Gcaba as taxi owner and nephew of Jacob Zuma, see Agiza Hlongwane, "Zuma's Nephew in R300m Tender Dispute," IOL News, December 9, 2012, http://www.iol.co.za/news/south-africa/kwazulu-natal/zuma-s-nephew-in-r300m-tender-dispute-1.1437844#.UXA21b_Xf0A.

11 *Brad's new employers:* South African Press Association, "Aurora to Pay R10m—Court," *Mining Weekly*, January 12, 2012, http://www

.miningweekly.com/article/aurora-to-pay-r10m-court-2012
-01-12; Sarah Britten, "Khulubuse Zuma's Lifestyle Thrust into
Auction," *Mail & Guardian*, April 24, 2012, http://www.mg.co
.za/article/2012-04-24-khulubuse-zumas-lifestyle-thrust-into
-auction; "Zondwa Mandela Faces Charges over Aurora," *Mail
& Guardian*, December 11, 2011, http://www.mg.co.za/article/
2011-12-11-zondwa-mandela-faces-charges-over-aurora; Martin
Plaut, "Mandela and Zuma Goldmine 'Exploiting workers,' " BBC
News, May 5, 2011, http://www.bbc.co.uk/news/world-africa-132
75704.

12 *The town of Springs:* H. E. Frimmel, D. I. Groves, et al., "The Formation and Preservation of the Witwatersrand Goldfields, the
World's Largest Gold Province," in *Economic Geology*, 100th An-
niversary volume, 2005, 769–97, http://www.geo.arizona.edu/geo
6xx/geo646a/646A_PW/Papers/Surface-related_Papers/Frimmel
05_WitsAu_EG100thAV.pdf.

12 *Forty percent of all the gold:* Ibid.

12 *The property covered:* Herbie Trouw, Aurora underground manager
and thirty-year veteran of the goldfield, interview with author.

12 *Yet underground, the Aurora mine:* Ibid.

15 *In the exchange of fire*: See also Shain Germaner, " 'Bad Brad' Group
Had Been Scouting," IOL News, September 14, 2011, http://
www.iol.co.za/news/south-africa/mpumalanga/bad-brad-group
-had-been-scouting-1.1137200#.UXA74b_Xf0A; "Aurora Justifies
Mine Killings," *Mail & Guardian*, August 13, 2010, http://mg.co
.za/article/2010-08-13-aurora-justifies-mine-killings; "Mystery of
Aurora Corpses," *Mail & Guardian*, August 13, 2010, http://mg.co
.za/article/2010-08-13-mystery-of-aurora-corpses.

18 *A 2001 monograph:* Peter Gastrow, *Theft from South African Mines
and Refineries*, Chapter 2, "The Product Theft of Gold," Monograph No. 54, 2001, Institute for Security Studies, http://www.iss
.co.za/Pubs/Monographs/No54/Chap2.html.

20 *At the Barberton mine:* Martin Creamer, "Barberton's Criminal Mining Smashed," *Mining Weekly*, September 10, 2010, http://www.miningweekly.com/article/barbertons-criminal-mining smashed.

21 *At just one of its mines:* Gold Fields production from http://www.goldfields.co.za/ops_south_deep.php; gold price from Kitco at http://www.kitco.com/scripts/hist_charts/yearly_graphs.plx.

21 *How many companies:* For Gold Fields fourth-quarter profit in 2011 see http://www.iol.co.za/business/companies/gold-fields-profits-increase-1.1236780#.Ubmm2JXXf0A.

22 *I have a little catalogue: Thracian Treasures* (Varna, Bulgaria: Slavena Publishing House, 2006).

22 *Today our asset menu:* For date of invention of money, see Peter Bernstein, *The Power of Gold: The History of an Obsession* (New York: John Wiley & Sons, 2000), 32.

22 *Yet by the fourteenth century:* Ibid., 110.

CHAPTER 2: RIVER OF GOLD

PAGE

25 *Spaniards came well equipped:* See for example http://www.classical fencing.com/horsetraining.php; http://www.pbs.org/gunsgerms steel/pdf/episode2.pdf.

25 *When he set out:* Bernstein, *The Power of Gold*, 113.

26 *He mentioned it 114 times:* See Jennifer Marx, *The Magic of Gold* (New York: Doubleday, 1978), 322–24; Richard Cowen at mygeology page.ucdavis.edu/cowen/~GEL115/115ch8.html. I have preferred Cowen's number as a more recent count.

26 *"A thing like a ball of stone":* http://www.clio.missouristate.edu/chuchiak/hst%20350-theme%209-spanish-weapons-and-armor .htm.

26 *"Gazing on such wonderful sights":* http://www.chnm.gmu.edu/worldhistorysources/sources/conquestofnewspain.htm.

26 *The Florentine Codex:* Fray Bernardino de Sahagún, "Malinche Begs Mexicas to Help Spaniards," *Florentine Codex*, Book 12, Chapter 18, http://www.faculty.fullerton.edu/nfitch/nehaha/aztec11.html; also at http://www.historians.org.

27 *He was the bastard son:* Bernstein, *The Power of Gold,* 122–23.

28 *The culture and appearance of the Andean people:* John Hemming, *The Conquest of the Incas* (London: Macmillan, 1970; rev. ed., Penguin, 1983), 60.

28 *They worked as part of:* María Rostworowski de Diez Canseco, *History of the Inca Realm* (Cambridge: Cambridge University Press, 1999), 182.

28 *They could move their armies:* Liesl Clark, "The Lost Inca Empire," January 11, 2000, http://www.pbs.org/wgbh/nova/ancient/lost-inca-empire.html.

28 *"Such magnificent roads":* Hemming, *The Conquest of the Incas*, 101.

29 *Their surgeons could drill holes:* Valerie A. Andrushko and John W. Verano, "Prehistoric Trepanation in the Cuzco Region of Peru: A View into an Ancient Andean Practice," *American Journal of Physical Anthropology* 137 (2008): 4–13.

29 *Early in 1527:* Hemming, *The Conquest of the Incas,* 25.

29 *They had "glimpsed the edges of a great civilization":* Ibid., 26.

30 *The Inca's ancestors:* Ibid., 121.

30 *The throne did not pass:* Ibid., 29.

31 *He did not know whether his forces:* Ibid., 30.

31 *Now the Spanish force entered the shadow:* Account of Spanish march to Cajamarca, first meeting with Atahualpa, and all events up to attack on Atahualpa and slaughter of Incas: Ibid., 31–43.

36 *He had meant to capture them:* Ibid., 45.

36 *With the Inca in Spanish hands:* Ibid., 49.

36 *They valued gold:* That gold was not the most valuable substance: from Ben Roberts, formerly head of the Department of Prehistory, British Museum, London, now on the faculty of Durham University.

36 *In Cajamarca today:* Hemming, *The Conquest of the Incas,* 48.

37 *A solid-gold sacrificial altar:* Ibid., 65. Hemming says it weighed 19,000 pesos, and I have converted using 1 peso = 27.5 grams.

37 *The wonders dazzled the Spaniards:* Asian influence; see Bernstein, *The Power of Gold,* 129.

37 *In a single month they retrieved:* Ibid., 130; conversion rate from http://www.measuringworth.com.

37 *In Atahualpa, Pizarro had the tap:* Hemming, *The Conquest of the Incas,* 50–51.

37 *Because of the Inca's divinity:* Inca ruling through generals and Spaniards stripping Cuzco: Ibid., 55–65.

38 *On April 14, 1533:* Atahualpa's last days: Ibid., 71–78.

40 *Pizarro was vilified:* Ibid., 80–83.

40 *Some died in fights:* Pizarro's death: Bernstein, *The Power of Gold,* 131.

CHAPTER 3: THE MASTER OF MEN

PAGE

41 *Spanish galleons:* See general history in "The Spanish Treasure Fleets of 1715 and 1733: Disasters Strike at Sea," http://www.nps.gov/nr/twhp/wwwlps/lessons/129shipwrecks/.

41 *In 1523 the French corsair:* http://www.time.com/time/specials/packages/article/0,28804,1860715_1860714_1860704,00.html.

42 *There would be "no peace beyond the line":* Ian K. Steele, "Imperial Wars," in *The Oxford Companion to American Military History*, ed. John Whiteclay Chambers II (New York: Oxford University Press, 1999), 327.

42 *Even so the great fleets:* Bernstein, *The Power of Gold*, 135.

42 *Spain had driven out:* Expulsion of Jews and Muslims: Bernstein, *The Power of Gold*, 140.

43 *"Gold and silver merely acquired":* Marie-Thérèse Boyer-Xambeu, Ghislain Delaplace, and Lucien Gillard, *Private Money and Public Currencies: The 16th Century Challenge*, trans. Azizeh Azodi (Armonk, NY: M. E. Sharpe, 1994), 116.

43 *Merchants increased the use of bills of exchange:* Bernstein, *The Power of Gold*, 153–57.

43 *In the late 1600s:* Stephen Quinn, "Goldsmith-Banking: Mutual Acceptance and Interbanker Clearing in Restoration London," *Explorations in Economic History* 34 (1997), 411–32, http://www.econ.tcu.edu/quinn/finhist/readings/goldsmiths.pdf.

44 *The great historian:* Barry Eichengreen, *Golden Fetters: The Gold Standard and the Great Depression, 1919–1939* (New York: Oxford University Press, 1996).

44 *The president of the World Bank:* Alan Beattie, "Zoellick Seeks Gold Standard Debate," *Financial Times*, November 7, 2010, http://ft.com/cms/s/0/eda8f512-eaae-11df-b28d-00144feab49a.html#axzz2OfGvMLkz.

44 *On the American right:* See for example Ron Paul, "Honest Money," http://www.ronpaul.com/on-the-issues/fiat-money-inflation-federal-reserve-2/: "If our money were backed by gold and silver,

people couldn't just sit in some fancy building and push a button to create new money. They would have to engage in honest trade with another party that already has some gold in their possession."

44 *Similar silver coins:* Bernstein, *The Power of Gold,* 78.

45 *In 1785:* see for example "Thomas Jefferson, Propositions respecting Coinage," *The Founders' Constitution,* University of Chicago Press web edition, eds. Philip B. Kurland and Ralph Lerner, http://press-pubs.uchicago.edu/founders/documents/al_8_5s5.html.

45 *Congress set a silver standard:* See the Coinage Act of April 2, 1792, http://www.constitution.org/uslaw/coinage1792.txt.

45 *A declining world gold supply:* Bernstein, *The Power of Gold,* 247.

45 *In 1797 the report of a French fleet:* Ibid., 199–201.

45 *In 1803 they sold Louisiana:* See first paragraph of note to "Primary Documents in American History, Louisiana Purchase," online at Library of Congress, http://www.loc.gov/rr/program/bib/ourdocs/Louisiana.html.

46 *The bankers running the transaction:* For details of the sale, see http://www.baringarchive.org.uk/features_exhibitions/louisiana_purchase.

46 *By 1834 the supply of gold in America:* Congress reset gold price: Bernstein, *The Power of Gold,* 248.

47 *On January 24, 1848, James Marshall panned some bright flakes:* Timothy Green, *The New World of Gold* (New York: Walker, 1981), 5.

47 *In the next seven years, 500,000 men:* "The California Gold Rush," http://www.ceres.ca.gov/ceres/calweb/geology/goldrush.html.

47 *"The whole country from San Francisco to Los Angeles":* Green, *The New World of Gold,* 5.

47 *"Three men using nothing but spoons":* Richard B. Lyttle, *The Golden*

Path: The Lure of Gold Through History (New York: Atheneum, 1983), 110.

47 *Dozens of new companies formed:* Maureen A. Jung, "Capitalism Comes to the Diggings," in *A Golden State: Mining and Economic Development in Gold Rush California*, eds. James J. Rawls and Richard J. Orsi (Berkeley: University of California Press, 1999), 53–54.

47 *The inrush of investment supported the development:* Robert Whaples, "California Gold Rush," http://www.eh.net/encyclopedia/article/whaples.goldrush.

48 *The chain pumps brought to California:* Ronald H. Limbaugh, "Making Old Tools Work Better," in *A Golden State*, 31.

48 *Tradition bathes the California gold rush in a honeyed light:* Daniel Cornford, " 'We All Live More like Brutes than Humans': Labor and Capital in the Gold Rush," in *A Golden State*, 78.

48 *Some Americans brought slaves:* Ibid., 84.

48 *A miner had to wash 160 pails:* Freezing water and brutish work: Ibid., 89.

48 *Even the investors suffered:* Jung, "Capitalism Comes to the Diggings," in *A Golden State*, 54.

49 *In terms of today's money:* Measuringworth.com equates purchasing power of $500 million in 1848 equal to $15 billion today.

49 *Robert Whaples:* Email from Robert Whaples to author, March 4, 2013.

49 *At the official U.S. government price:* Gold production nearly 2 percent of U.S. GDP: Whaples, "California Gold Rush."

49 *New discoveries elsewhere in the world added even more production:* Green, *The New World of Gold*, 1.

49 *"As the creditor of the whole earth":* Growth in Bank of England and Bank of France gold stocks: Ibid., 9.

49 *In 1871 Germany bought £50 million worth of gold:* Ibid., 21.

49 *"We chose gold":* Bernstein, *The Power of Gold,* 250.

49 *The silver dominoes began to fall:* Green, *The New World of Gold,* 21.

50 *But some scholars think:* See especially Marc Flandreau, *The Glitter of Gold: France, Bimetallism, and the Emergence of the International Gold Standard, 1848–1873* (New York: Oxford University Press, 2004).

50 *In 1861, when the costs:* Bernstein, *The Power of Gold,* 262.

50 *As the cost of wheat shot up in Europe, America had a bumper crop:* Ibid., 268.

50 *In 1890, a banking crisis in Argentina:* Ibid., 252–53, 268.

51 *The havoc that followed:* Ibid., 270–72.

51 *The government had $40 million:* Ibid., 272.

51 *In American terms, the Morgans had been rich forever:* Ron Chernow, *The House of Morgan: An American Banking Dynasty and the Rise of Modern Finance* (New York: Grove Press, 1990), 17ff.

53 *As the sense of crisis heightened:* Ibid., 74 ff.

54 *For all its harshness:* Ibid., 74ff.

54 *The system solved two main problems:* Author interview with Professor Michael Woodford, Columbia University. (In case there is any doubt about Professor Woodford's position: he does not at all support a return to the gold standard.)

55 *Barry Eichengreen:* "A Critique of Pure Gold," *National Interest,* August 24, 2011, http://www.nationalinterest.org/article/critique-pure-gold-5741.

56 *It had 20,000 tons of bullion:* Timothy Green, *Central Bank Gold Reserves: An Historical Perspective Since 1845* (London: World Gold Council, 1999), 18, http://www.gold.org/download/pub_archive/pdf/Rs23.pdf.

CHAPTER 4: CAMP DAVID COUP

PAGE

57 *Gold is surging out of the Bank of England:* Gold movements and decrease of U.S. gold stock: Bernstein, *The Power of Gold,* 311–23.

58 *On April 5 he signed Executive Order 6102:* For the text of the order, see http://www.presidency.ucsb.edu/ws/index.php?pid=14611.

59 *A challenger immediately tested the law:* "Lawyer Indicted as Gold Hoarder," *New York Times,* October 6, 1933, http://query.nytimes.com/mem/archive/pdf?res=FA0815FB3E541A7A93C4A9178B D95F478385F9; "Frederick Campbell, Lawyer, Is Dead Here," *New York Times,* December 27, 1937, http://query.nytimes.com/mem/archive/pdf?res=F60D14F93C5D12738DDDAE0A94DA 415B878FF1D3.

59 *In 1930 the United States:* Green, *Central Bank Gold Reserves,* 17.

60 *As soon as he had issued the gold confiscation order, Roosevelt obtained the authority:* Roosevelt and Morgenthau rigging price: Bernstein, *The Power of Gold,* 321–22.

61 *"We fight together on sodden battlefields":* http://www.centerfor financialstability.org/brettonwoods.php.

61 *A newly discovered transcript:* Annie Lowry, "Transcript of 1944 Bretton Woods Conference Found at Treasury," *New York Times,* October 25, 2012, http://www.nytimes.com/2012/10/26/business/transcript-of-1944-bretton-woods-meeting-found-at-treasury .html.

61 *"Now the advantage is ours here":* Fred Andrews, "A Grudge Match for Global Finance," review of Benn Sterl, *The Battle of Bretton Woods, New York Times,* March 2, 2013, http://nytimes.com/2013/03/03/business/bretton-woods-monetary-agreement-examined-in-a-new-book.html.

62 *For a while, that's how it did work, with Japan and Europe, "eager for dollars they could spend":* How U.S. share of world economy dropped from 35 percent to 27 percent: Roger Lowenstein, "The Nixon Shock," *Bloomberg Businessweek,* August 4, 2011, http://www.businessweek.com/magazine/the-nixon-shock-08042011.html.

62 *Surplus dollar holdings:* Bernstein, *The Power of Gold,* 334.

63 *In a span of only thirteen years:* U.S. finances deteriorate: Bernstein, citing Jacques Rueff, ibid., 338–39.

63 *"Gold!" he intoned, the standard that "has no nationality":* Berstein, *The Power of Gold,* 329.

63 *Two years later:* De Gaulle pulls out of Gold Pool: Ibid., 339–41.

64 *On March 11, 1968, only days after the huge bullion transfer from the United States, the pool vowed to hold the line at $35:* Clyde H. Farnsworth, "U.S. Wins Backing from Gold Pool to Hold $35 Price," *New York Times,* March 11, 1968, http://query.nytimes.com/mem/archive/pdf?res=F30E11FE3A55157493C3A81788D85F4C8685F9.

64 *A week later the pool folded:* Edwin L. Dale, Jr., "7 Nations Back Dual Gold Price, Bar Selling to Private Buyers; Pledge Support of the Dollar," *New York Times,* March 18, 1968, http://query.nytimes.com/mem/archive/pdf?res=FA0D14F93E541A7493CAA81788D85F4C8685F9.

66 *Connally's biographer believed that the assassination enhanced Connally's reputation:* Charles Ashman, *Connally: The Adventures of Big Bad John* (New York: Morrow, 1974), 15.

67 *Herbert Stein, who knew Connally and saw him in operation:* Description of Connally as "forceful, colorful, charming": Herbert Stein, *Presidential Economics: The Making of Economic Policy from Roosevelt to Reagan and Beyond* (New York: Simon & Schuster, 1985), 162.

67 *"I can play it round or I can play it flat":* Ibid., 163.

67 *One of Nixon's closest aides:* "Connally Awed Nixon, as Haldeman Diaries Tell," *Baltimore Sun,* May 20, 1994, http://articles .baltimoresun.com/1994-05-20/news/1994140195_1_connally -nixon-new-party.

68 *Crucial to the success of the two-step plan was secrecy:* Stein, *Presidential Economics,* 166.

69 *Two days later:* H. Erich Heinemann, "Speculative Attacks Grow on U.S. Currency Abroad; Nervous Foreign Dealers Sell Dollars as Federal Reserve Bolsters Monetary Defenses with a Swap Deal," *New York Times,* August 13, 1971, http://query.nytimes.com/mem/archive/ pdf?res=F40C13F9395C1A7493C1A81783D85F458785F9.

70 *An envoy from the Bank of England:* Stein, *Presidential Economics,* 167.

70 *Then Connally took over the meeting:* Lewis E. Lehrman, "The Nixon Shock Heard 'Round the World," *Wall Street Journal,* August 15, 2011, http://online.wsj.com/article/SB100014240531119040073 04576494073418802358.html.

71 *The retreat had had a storied place in American affairs:* See for example entry on Camp David at http://www.nps.gov.

72 *Calling the situation a "ferment":* Heinemann, "Speculative Attacks Grow on U.S. Currency Abroad."

73 *On the brink of dismantling the global financial system, he worried about interrupting the hit TV western* Bonanza: Daniel Yergin and Joseph Stanislaw, *The Commanding Heights: The Battle Between*

Government and the Marketplace That Is Remaking the Modern World (New York: Simon & Schuster, 1998), 62.

73 *President Lyndon Johnson had made it policy:* Edward Hudson, "Lorne Greene, TV Patriarch, Is Dead," *New York Times*, September 12, 1987, http://www.nytimes.com/1987/09/12/arts/lorne -greene-tv-patriarch-is-dead.html.

73 *"The P. was down in his study with the lights off and the fire going":* Yergin and Stanislaw, *The Commanding Heights,* 63.

73 *One dealer characterized the trading as more like a flea market:* "Tailspin in Dollar Sector of Eurobonds Is Severe," *New York Times,* August 16, 1971, http://query.nytimes.com/mem/archive/pdf?res= FB0D10FD3D591A7493C4A81783D85F458785F9.

74 *"In the past seven years":* Text of Nixon Shock speech available at the American Presidency Project, University of California, Santa Barbara, http://www.presidency.ucsb.edu/ws/index.php?pid=3115 &st=&st1.

74 *The first market with a chance to react was Tokyo:* "Exchanges in Tokyo Unloading Dollars," *New York Times*, August 16, 1971, http://query.nytimes.com/mem/archive/pdf?res=F70A12FD3 D591A7493C4A81783D85F458785F9.

74 *As dismay spread around the globe, the White House awaited the American response:* Stein, *Presidential Economics,* 180.

75 *"When you wake up in the morning, do you care about the price of gold?":* Floyd Norris, "In a Focus on Gold, History Repeats Itself," *New York Times*, February 2, 2012, http://www.nytimes.com/ 2012/02/03/business/in-rise-of-gold-bugs-history-repeats-itself .html?pagewanted=all.

CHAPTER 5: THE DISCOVERY OF INVISIBLE GOLD

PAGE

77 *"While I was working in the field":* Ralph J. Roberts, *A Passion for Gold: An Autobiography* (Reno: University of Nevada Press, 2002), 28.

77 *"On those trips into the mountains":* Ibid., 47.

78 *He stepped off the bus in Winnemucca:* Ibid., 29.

78 *There was a working gold mine in the heart of Willow Creek:* Ibid., 34.

78 *Mining in Nevada had a long pedigree:* "Outline of Nevada Mining History," Nevada Bureau of Mines and Geology, quarterly newsletter, Fall 1993, http://www.nbmg.unr.edu/dox/nl/n120.htm.

79 *To explain what had happened, Roberts hypothesized:* Roberts, *A Passion for Gold,* 34.

80 *In 1939, when Roberts began his explorations:* Ibid., 40–43.

80 *The writers described huge tables of Triassic rock:* Arnold Hague and S. F. Emmons, "Descriptive Geology," in *United States Geological Exploration of the Fortieth Parallel,* ed. Clarence King (Washington, DC: U.S. Government Printing Office, 1877), 601. The volumes have been digitized by the Boston Public Library and are available online. The geology volume can be accessed at http://www.archive.org/stream/descriptivegeolo00hagu#page/n11/mode/2up.

80 *He struck out along the Antler range in his happiest state—alone:* Roberts, *A Passion for Gold,* 41.

80 *When the ocean plate and continental plate collided:* Dean G. Heitt, "Newmont's Reserve History of the Carlin Trend, 1965–2001," in *Gold Deposits of the Carlin Trend,* eds. Tommy B. Thompson, Lewis Teal, and Richard O. Meeuwig (Reno: Nevada Bureau of Mines and

Geology Bulletin 111, University of Nevada), 2002, 35ff, http://www.nbmg.unr.edu/dox/b111/history.pdf.

81 *"As I approached the western margin of the range"*: Roberts, *A Passion for Gold,* 44.

82 *Ralph Jackson Roberts was born in 1911:* Ibid., 12–19.

82 *Even the Roberts Mountains of north-central Nevada:* Ibid., 83.

83 *"Our eyes met and held for a long moment":* Roberts's wife, Arleda: Ibid., 28–29, 51–53.

83 *He saw that the feature called the Roberts Mountains thrust presented a continuous geological relationship:* Ibid., 84.

84 *William Vanderburg:* J. Alan Coope, *Carlin Trend Exploration History: Discovery of the Carlin Deposit* (Reno: Nevada Bureau of Mines and Geology Special Publication 13, University of Nevada, 1991), 5, http://www.nbmg.unr.edu/dox/sp13.pdf.

84 *"Bill wanted to show me this unusual ore":* Roberts, *A Passion for Gold,* 86.

85 *"When we entered Maggie Creek Canyon":* Ibid., 87.

85 *"My excitement grew":* Ibid., 88.

86 *In 1960 Roberts put his thoughts:* "Alinement of Mining Districts in North-Central Nevada," United States Geological Survey Professional Paper 400-B, 1960, B17–B19.

87 *One man stayed behind:* John Seabrook, "A Reporter at Large: Invisible Gold," *New Yorker,* April 24, 1989, 69–70.

88 *John Sealy Livermore:* Ibid., 62. See also "John Sealy Livermore, 1918–2013," obituary in *Napa Valley Register,* February 14, 2013, http://napavalleyregister.com/obituaries/john-sealy-livermore/article_3048b3d8-7646-11e2-9e18-0019bb2963f4.html.

88 *Livermore's search for invisible gold began in 1949 at the Standard mine:* Seabrook, "A Reporter at Large: Invisible Gold," 65.

88 *"We did some churn drilling":* John Sealy Livermore, "Prospector, Geologist, Public Resource Advocate: Carlin Mine Discovery, 1961; Nevada Gold Rush, 1970s," *Western Mining in the Twentieth Century Oral History Series* (Berkeley: University of California, 2001), 29; navigate from http://bancroft.berkeley.edu/ROHO/projects/mining/.

89 *In Livermore's conjecture:* Seabrook, "A Reporter at Large: Invisible Gold," 65.

90 *"I get lonely sometimes, too":* Ibid., 68.

90 *"That really got me interested":* Livermore, oral history, 66.

91 *Then Livermore visited Harry Bishop:* Ibid., 68.

92 *"My God, it was a beautiful ore body":* Seabrook, "A Reporter at Large: Invisible Gold," 73.

93 *We don't know a thing about him except where he slept his hopes away:* Biggest goldfield in America: Lewis Teal and Mac Jackson, "Geologic Overview of the Carlin Trend Gold Deposits," in *Gold Deposits of the Carlin Trend*, 9.

CHAPTER 6: GOLDSTRIKE!

PAGE

96 *Three years later the gold price passed $180:* For historic gold prices, see http://www.kitco.com/charts/historicalgold.html; http://www.kitco.com/gold.londonfix.html.

96 *"Nobody in his right mind":* "Gold Surges Again in a 'Buying Fever,' " Associated Press, December 28, 1979, http://query.nytimes.com/mem/archive/pdf?res=F00A1EF73F5410728DDDA10A94DA415B898BF1D3.

97 *In 1944, when the German wartime occupying force:* Early Munk family details: Anna Porter, *Kasztner's Train: The True Story of Rezso Kasztner, Unknown Hero of the Holocaust* (Toronto: Douglas & McIntyre, 2007), 222ff.

100 *Six years later he and business partner David Gilmour founded Clairtone:* See for example Nina Munk and Rachel Gotlieb, *The Art of Clairtone: The Making of a Design Icon, 1958–1971* (Toronto: McClelland & Stewart, 2008); http://www.clairtone.ca/projectg/.

100 *A 1967 commercial posted on YouTube:* http://www.youtube.com/watch?v=K7wiQ1At_9Q.

100 *The Clairtone company was falling apart around them:* Nina Munk, "The Lesson of Clairtone," *Canadian Business,* May 12, 2008, 42; also http://www.ninamunk.com/documents/ClairtoneLesson.htm, under the heading "My Father's Brilliant Mistake."

101 *"Munk was too good a salesman for his own good":* Ibid.

101 *"Desperate for capital":* Munk's hotels, and interview with Lord Inchcape: Michael Posner, "Peter Munk's Reflections on Being a Winner," *Globe and Mail,* Toronto, February 18, 2011, http://www.theglobeandmail.com/news/national/peter-munks-reflections-on-being-a-winner/article567172/?page=all.

103 *In one biography he arrives for his deportation from Hungary brimming with unconcern:* Richard Rohmer, *Golden Phoenix: The Biography of Peter Munk* (Toronto: Key Porter Books, 1997), 9.

103 *Another chronicle contains a well-worn story about Munk:* Posner, "Peter Munk's Reflections on Being a Winner."

103 *He built the biggest gold miner in the world:* See for example Liezel Hill, "Barrick to Lead Wave of Gold-Mining Asset Sales in 2013," *Bloomberg,* January 8, 2013, http://www.bloomberg.com/news/2013-01-08/barrick-seen-leading-gold-asset-sales-in-2013-corporate-canada.html, in which Barrick is called "the larg-

est producer of the precious metal"; Trefis Team, "Barrick Gold Can Still Sparkle Despite Higher Costs and Impairment Charges in Results," *Forbes*, February 19, 2013, http://www.forbes.com/sites/greatspeculations/2013/02/19/barrick-gold-can-still-sparkle-despite-higher-costs-and-impairment-charges-in-results/, where Barrick is called "the largest gold producer in the world."

103 *Barrick bought its first mine in 1984:* Account of Barrick's growth from interviews with Peter Munk and Alan Hill, and from Peter C. Newman, *Dreams and Rewards: The Barrick Story* (Toronto: Barrick Gold, 1996).

103 *In raising the money, Munk showed the financial audacity:* See "Jolly Gold Giant: How Peter Munk Built the World's Biggest Gold Miner from Scratch in Just 25 Years," *Economist*, April 17, 2008, http://www.economist.com/node/11045502.

112 *Goldstrike powered the company into the ranks of the world's biggest gold producers:* For the success of Goldstrike, see for example Clyde H. Farnsworth, "Profile/Peter Munk; Lucky Gold Strike and Wise Hedging Help Keep His Shareholders Smiling," *New York Times*, May 30, 1993, http://www.nytimes.com/1993/05/30/business/profile-peter-munk-lucky-gold-strike-wise-hedging-help-keep-his-shareholders.html?pagewanted=all&src=pm.

113 *In forward selling, a miner contracts with a commercial bank:* The mechanics of forward selling were explained to me by Peter Munk and by Richard Young. Young is president and CEO of Teranga Gold Corporation, and a veteran of gold mine finance and of Barrick.

113 *A single input can demolish the forward seller's math:* "Barrick to Sell $3 Billion in Stock to Buy Back Hedges," Reuters, September 8, 2009, http://www.reuters.com/article/2009/09/08/us-barrick-hedge-idUSTRE58767F20090908.

CHAPTER 7: LINGLONG

PAGE

115 *In a ten-year exploration binge:* See for example Maria Kolesnikova, "Top 10 Gold-Producing Countries in 2011," *Bloomberg,* April 5, 2012, http://www.bloomberg.com/news/2012-04-05/top-10-gold -producing-countries-in-2011-table.html; for China replacing South Africa in 2007: Shu-Ching Jean Chen, "China Became World's Top Gold Producer In 2007," *Forbes,* January 18, 2008, http://www.forbes.com/2008/01/18/china-gold-production -markets-econ-cx_jc_0118markets02.html.

115 *They were buying gold too:* "China to Overtake India in Overall Gold Demand: GFMS," Reuters, November 8, 2012, http://in.reuters .com/article/2012/11/08/gold-china-gfms-india-idINDEE8A 706R20121108.

116 *All together, they produce more than 300 tons:* Kolesnikova, "Top 10 Gold-Producing Countries in 2011."

116 *My attention settled on a picture:* Yellow book: Gong Runtan and Zhu Fengsan, *The Gold Mining History of Zhaoyuan with a Review of the Gold Industry in the P.R.* [People's Republic of] *China,* editor-in-chief, Wei Youzhi (China Ocean Press and Placer Dome China Ltd., 2004), 4.

118 *In ancient times, the legend says:* Ibid., 53–54.

118 *The Chinese were mining gold by 1300 BC:* And cinnabar as indicator: Ibid., 2–7.

119 *"If in the mountain grow spring shallots, there will be silver under the ground; if leek in the mountain, gold.":* Ibid., 34.

119 *In 1980 researchers at the University of London:* Christine A. Girling and Peter J. Peterson, "Gold in Plants," *Gold Bulletin* 13, no. 4 (December 1980): 151–57, http://www.link.springer.com/ article/10.1007/BF03215461?LI=true.

119 *A 1998 article in* Nature: Christopher W. N. Anderson, Robert R. Brooks, Robert B. Stewart, and Robyn Simcock, "Harvesting a Crop of Gold in Plants," *Nature,* October 8, 1998, No. 395, 553–54, http://www.nature.com/nature/journal/v395/n6702/full/395553a0.html.

119 *In 2004 a New Zealand researcher:* Jen Ross, "Money That Grows on Crops," *Christian Science Monitor,* April 14, 2004, http://www.csmonitor.com/2004/0415/p17s02-sten.html.

119 *And a paper published:* Victor Wilson-Corral, Mayra Rodriguez-Lopez, Joel Lopez-Perez, Miguel Arenas-Vargas, and Christopher Anderson, "Gold Phytomining in Arid and Semiarid Soils," paper at the World Congress of Soil Science, August 1–6, 2010, Brisbane, Australia, http://www.iuss.org/19th%20WCSS/Symposium/pdf/1593.pdf.

119 *They had novel ways of refining gold:* Gong and Zhu, *The Gold Mining History of Zhaoyuan,* 43.

119 *At the beginning of the first millennium:* Gold reserves equal to those of Rome: Ibid., 3.

119 *When the Mongols came to power:* Ibid., 6.

120 *In the seventeenth century Linglong came into the hands of the famous eunuch Wei Zhongxian:* Ibid., 7. See also "Wei Zhongxian—The Most Powerful and Notorious Eunuch in Chinese History," *Cultural China,* http://history.cultural-china.com/en/47History6380.html; "Wei Zhongxian," *Encyclopaedia Britannica,* http://www.britannica.com/EBchecked/topic/638872/Wei-Zhongxian.

120 *The new Manchu rulers hated gold mining:* Gong and Zhu, *The Gold Mining History of Zhaoyuan,* 10.

121 *By 1888 China was producing 700,000 ounces a year:* Ibid., 11. Gong and Zhu use the figure 432,000 liang. I used two online converters: http://www.convert-me.com/en/convert/weight/cliang

.html; http://www.convertcenter.com/chinese-liang. Each gave a metric weight of 21.6 million grams, which I divided by 31 to convert to troy, finding 696,774 ounces.

122 *"The accumulated snow on Mount Linglong had not yet melted":* Account of Japanese occupation: Gong and Zhu, *The Gold Mining History of Zhaoyuan,* 64ff.

124 *Approved by Mao Zedong to root out what he saw as capitalist sympathies:* See for example Roderick MacFarquhar and Michael Schoenhels, "Red Terror," Chapter 7, *Mao's Last Revolution* (Cambridge: Harvard University Press, 2006), 117–31.

126 *Spurred by the prospect of potential gains:* Establishment of the Gold Army: Gong and Zhu, *The Gold Mining History of Zhaoyuan,* 91.

126 *Zhu became adviser to the top Gold Army leaders:* Details of Zhu's rise from Zhu Fengsan and Greg Hall, an Australian consultant geologist, and friend of Zhu's. Hall is a director of four gold mining companies, including China Gold International Resources Corp. Ltd., whose majority shareholder is state-owned China National Gold Group. He was an executive of the Australian subsidiary of Placer Dome, a large miner (later bought by Barrick) that entered China in 1997 to look for a gold property.

CHAPTER 8: THE BANDIT CIRCUS

PAGE

129 *In 1993, to encourage exploration:* Greg Hall and Zhu Fengsan interviews with author. But see also William L. MacBride, Jr., and Wang Bei, "Chinese Mining Law Overview," available as a pdf from Gough, Shanahan, Johnson, and Waterman, a Montana natural resources law firm. The paper says mining law had not changed until 1996, but mentions "regulations being promulgated in 1993 and 1994," http://www.gsjw.com/catalogs/catalog110/section174/

file51.pdf. In a March 13, 2013, email to author, MacBride speculated that the 1993 regulations were China's signal to the West that mining investment would be encouraged in the rewrite of the mining laws that China was then drafting.

129 *In 1997 China was the world's fifth-ranked gold producer:* United States Geological Survey, "Gold Statistics and Information," http://minerals.usgs.gov/minerals/pubs/commodity/gold/.

130 *Placer Dome had a five-year plan of its own:* Greg Hall, interview with author. See also slides of Hall's presentation to Prospectors and Developers Association of Canada conference, 2003, http://www.pdac.ca/pdf-viewer?doc=/docs/default-source/publications-papers-presentations-conventions/hall.pdf.

130 *It was a pleasure the Chinese were anxious to supply to others:* Account of Mundoro experience from Colin McAleenan interview with author. I confirmed the details of McAleenan's version of events with an investment banker familiar with the story, who for reasons of business stipulated that I not name him. He said that, generally, Western gold miners had not done well in China. Email from richdad.com confirmed Robert Kiyosaki as an investor; Frank Crerie is now deceased.

133 *In November 2003 they took the company public and raised $13 million:* See company press release, December 4, 2003. (To view public documents such as press releases and material change reports for Canadian listed companies, go to http://www.sedar.com and navigate from the "Company Profiles" tab. The SEDAR system is maintained by the Canadian Securities Administrators, comprising regulators from every province and territory of Canada.)

133 *One of the most hyped of these was the Boka prospect:* See for example "Southwestern Resources Corp.: Positive Drill Results, Boka Gold Project, Yunnan Province, China," *Business Wire*, August 31, 2005, http://www.businesswire.com/news/home/20050831005839/en/Southwestern-Resources-Corp.-Positive-Drill-Results-Boka.

133 *Southwestern's market capitalization grew to more than three quarters of a billion dollars:* Brian Hutchinson, "Gold Mine's Results Transcended Belief," *National Post*, December 22, 2010. (The Internet link to the story has expired.)

133 *"The problem," said an official:* Shao Da, "Yunnan Gold Prospecting Project Held Up," China.org.cn, September 5, 2006, http://www.china.org.cn/english/2006/Sep/180144.htm.

133 *Southwestern's share price later collapsed:* Hutchinson, "Gold Mine's Results Transcended Belief"; and including Paterson sentence: David Baines, "Former Geologist Sentenced to Six Years in Prison for Assay Fraud," *Vancouver Sun,* January 18, 2013, http://www.vancouversun.com/business/David+Baines+Former+geologist+sentenced+years+prison+assay+fraud/7841963/story.html.

134 *They raised another $25 million:* See Mundoro Mining release, October 6, 2004, at http://www.sedar.com.

135 *Instead of 79 percent of profits:* See Mundoro Capital August 2, 2011, press release accepting deal, and calling it "the best current available alternative for shareholders of Mundoro to participate in the future development of the Maoling Gold Project," http:www.sedar.com.

136 *When China made the 1993 decision to encourage mine development, it allowed any Chinese company:* X. D. Jiang, vice president of production and executive director, China Gold International Resources Corp. Ltd., interview with author.

136 *Then it caught the eye of Robert Friedland:* Friedland as billionaire: "The World's Billionaires," *Forbes*. In March 2013, Friedland's net worth was said to be $1.8 billion: http:www.forbes.com/profile/robert-friedland/. Friedland as "Toxic Bob": see for example Scott Condon, " 'Toxic Bob' vs. Wealthy Neighbors," *Aspen Times,* July 16, 2009, http://www.aspentimes.com/article/20090716/NEWS/907159977; David Robertson, " 'Toxic Bob' Admits De-

feat in Power Struggle," *Times* (London), April 19, 2012, http://
www.thetimes.co.uk/tto/business/industries/naturalresources/
article3388506.ece; Joshua Zapf, "Robert Friedland," Mining.com,
June 29, 2012, http://www.mining.com/robert-toxic-bob-fried
land-49502/. Friedland cyanide spill: "Cyanide-Spill Suit Is Settled in Colorado," *New York Times,* December 24, 2000, http:www
.nytimes.com/2000/12/24/us/cyanide-spill-suit-is-settled-in-colo
rado.html.

136 *Chang Shan Hao presented a similar opportunity:* I relied on
X. D. Jiang for the chronology of Friedland/Jinshan involvement
at Chang Shan Hao, as he was intimate with the development from
the beginning. Jinshan press releases and material statements may
be viewed at the website of China Gold International, which took
over Jinshan: http://www.chinagoldintl.com, and enter "Jinshan"
into the search field.

137 *The year that China moved into first place:* See United States Geological Survey, Gold Statistics and Information, http://minerals.usgs
.gov/minerals/pubs/commodity/gold/.

137 *Gold became a restricted commodity:* X. D. Jiang and Greg Hall, interviews with author. See also for example "Foreign Investors' Gold
Rush in China Restricted," http://www.chinastakes.com/2008/4/
foreign-investors-gold-rush-in-china-restricted.html.

137 *Friedland was looking for money anyway:* Trish Saywell, "Ivanhoe
Sells Jinshan Stake," *Northern Miner,* April 21, 2008, http://north
ernminer.com/news/ivanhoe-sells-jinshan-stake/1000220684/.
See also "Ivanhoe Mines to Sell Controlling Stake in Jinshan Gold
Mines to China National Gold Group Corporation," joint press
release from parties, April 10, 2008, http://www.turquoisehill
.com/i/pdf/2008-04-10_NR.pdf.

137 *Eventually Friedland's joint venture partner in Mongolia:* Euan Rocha,
"Friedland Exits as Ivanhoe, Rio Sign Finance Deal," Reuters,
April 18, 2012, http://www.uk.reuters.com/article/2012/04/18/uk

-ivanhoe-rio-idUKBRE83H0WL20120418; Ian Austen, "Mining Chief Resigns as Part of Deal with Rio Tinto," *New York Times,* April 18, 2012, dealbook.nytimes.com/2012/04/18/mining-chief -resigns-as-part-of-deal-with-rio-tinto/.

138 *Inner Mongolia's fountain of coal:* Ben Woodhead, "Where China's Million Millionaires Live," *BRW* (*Business Review Weekly*), August 2012, http://www.brw.com.au/p/sections/eco/where_china_mil lion_millionaires_hlqEoC5O8nBfGHR8YNcMwJ; see also Kristine Lim, "Ordos Millionaires Fuelling Property Boom in Inner Mongolia," channelnewsasia.com, August 29, 2011, http://www .allvoices.com/news/10186482-ordos-millionaires-fuelling-prop erty-boom-in-inner-mongolia.

139 *A month before I visited in June of 2011, Mongol protests:* Andrew Jacobs, "Chinese Try to Appease Mongolians," *New York Times,* May 29, 2011, http://www.nytimes.com/2011/05/30/world/asia/ 30mongolia.html.

139 *China executed the truck driver:* "China Executes Truck Driver for Running Over Mongolian Herder," Associated Press, August 25, 2011, http://www.guardian.co.uk/world/2011/aug/25/china -truck-driver-executed-mongolian-herder.

141 *Chang Shan Hao had nineteen drill rigs:* X. D. Jiang, interview with author; mill expansion and production forecast from "China Gold International Resources Provides Preliminary 2012 Operating, Production and Exploration Highlights and 2013 Outlook," http://www.chinagoldintl.com/i/pdf/NR-Jan-29-2013.pdf.

142 *The mining contractor was China Railway:* For the size of China Railway, see for example Wu Zhong, "Blowing the Whistle on 'Big Brother,' " *Asia Times,* May 7, 2008, http://www.atimes.com/ atimes/China/JE07Ad01.html.

142 *Notoriously corrupt:* "Ex China Rail Minister Liu Zhijun Charged with Corruption," BBC News, April 10, 2013, http://www.bbc .co.uk/news/world-asia-china-22089011.

145 *In the 1980s, before China opened exploration to Western companies, the government tried to stimulate production:* Gong and Zhu, *The Gold Mining History of Zhaoyuan*, 91–92. Gong and Zhu cite "Circular About Key Points in Increasing Production of Gold and Silver," issued August 30, 1985, by the State Council, stating in part its purpose as "invigorating the gold economy and taking care of gold industry."

145 *Gold seekers ransacked prospective ground:* X. D. Jiang, interview with author; also Gong and Zhu, *The Gold Mining History of Zhaoyuan*, 94.

146 *In 1988 the Chinese government reversed its liberalizing policy:* Gong and Zhu, *The Gold Mining History of Zhaoyuan*, 102, 117.

146 *In 1989 the* Christian Science Monitor: James L. Tyson, "Farmers Drop Hoes to Dig for Gold," *Christian Science Monitor,* November 21, 1989, http://www.csmonitor.com/1989/1121/drush.html.

148 *In Henan province, 178 small-mining teams:* Gong and Zhu, *The Gold Mining History of Zhaoyuan,* 117.

148 *In an incident at the Wenyu mine:* Ibid., 118.

148 *Even large state companies used* min cai *to recover hard-to-get-at ore:* Ibid., 30.

148 *But in 2001, in another shift of central government sentiment:* Ibid., 31.

150 *In 1995 "gold lords" and their bands of peasant miners were still defying the government:* Patrick E. Tyler, "Gold Hunters, Defying Beijing, Mine Vast Areas of Rural China," *New York Times,* July 17, 1995, http://www.nytimes.com/1995/07/17/world/gold -hunters-defying-beijing-mine-vast-areas-of-rural-china.html?page wanted=all&src=pm.

151 *In 2012, nomadic Tibetan shepherds in Qinghai:* Marc Howe, "Tibetan Nomads Clash with Gold Miners over Sacred Mountain,"

Mining.com, August 13, 2012, http://www.mining.com/tibetan
-nomads-clash-with-chinese-gold-miners-over-sacred-mountain
-50519/.

151 *In 2010 the World Gold Council warned: China Gold Report: Gold
in the Year of the Tiger* (London: World Gold Council, 2010),
62–63.

152 *The China-running-out-of-gold scenario raises the specter of "peak
gold":* See for example Liezel Hill, "Iamgold Sees 'Peak Gold' Forg-
ing $2,500 Price," *Bloomberg,* January 17, 2013, http://www.bloom
berg.com/news/2013-01-17/iamgold-sees-peak-gold-forging-2
-500-price.html; Ambrose Evans-Pritchard, "Barrick Shuts Hedge
Book as World Gold Supply Runs Out," *Telegraph,* November 11,
2009, http://www.telegraph.co.uk/finance/newsbysector/industry/
mining/6546579/Barrick-shuts-hedge-book-as-world-gold-supply
-runs-out.html; Richard Kerr, "Is the World Tottering on the Prec-
ipice of Peak Gold?," *Science,* March 2, 2012, 1038–39, http:www
.sciencemag.org/content/335/6072/1038.

CHAPTER 9: THE SPIDER

PAGE

153 *The volume of bullion coming to market today:* United States Geological Survey, "Gold Statistics and Information," http://minerals
.usgs.gov/minerals/pubs/commodity/gold/, 2012 world production
2,700 tons. Gold today as multiple of Spanish imports in sixteenth
century: there are no reliable figures for world production, but here
is my back-of-the-envelope calculation. I took pre-sixteenth-century
world supply of 42 tons from Bernstein, *The Power of Gold* (2 cubic
meters). In the sixteenth century that supply grew by about five
times. 210 tons minus 42 tons = 168 tons for a century's production, or 1.68 tons a year. Divide 2,700 tons by 1.68 = 1,607 times
the earlier production.

153 *In 2011, the total value of mined gold was about $143 billion:* I took 2,700 metric tons production figure for 2011 from the United States Geological Survey, converted at 31 grams = 1 troy ounce, and used a mean gold price for 2011 of $1,650.

153 *In one three-month period alone, 11 billion ounces of gold worth $15 trillion changed hands in London:* Jack Farchy, "Sizing Up the Gold Market," *Financial Times,* September 9, 2011, http://www .ft.com/cms/s/0/eb342ad4-daba-11e0-a58b-00144feabdc0.html# axzz2NiR9WqWu.

153 *Most of the trade is in the hands of a cabal of banks in London:* See http://www.goldfixing.com, and also for example Robert D. Hershey, Jr., "Gold 'Fixing': London Tradition," *New York Times,* February 12, 1979, http://query.nytimes.com/mem/archive/pdf?res=FA0 E1EFD3A5413718DDDAB0994DA405B898BF1D3.

154 *The London Bullion Market Association polled its members:* See Farchy, "Sizing Up the Gold Market," but I checked the calculation: 11 billion troy ounces = 341 billion grams = 341,000 metric tons = 126 times 2011 production of 2,700 metric tons (from the United States Geological Survey) and twice the 170,000 metric tons in the world.

154 *The Spider is an exchange-traded fund:* Spider account from author's interviews with Jason Toussaint and George Milling-Stanley, executives of SPDR Gold Shares.

156 *There was more gold in the Spider's stash than in the central bank of China:* At the time, China's stated reserves were about 1,000 tons, although the central bank was thought to have more. See "China Could Soon Announce That Their Gold Reserves Have Doubled," *Business Insider,* January 22, 2013, http://www.businessinsider .com/russia-turkey-ukraine-buy-gold-but-bullion-tiny-part-of-fx -reserves-2012-8; "The World's Biggest Gold Reserves," CNBC, http://www.cnbc.com/id/33242464/page/10.

156 *Some of the owners' positions show up in 13F filings:* See for example "Soros Doubles Down on Gold," *New York Times,* February 2, 2010, dealbook.nytimes.com/2010/02/17/soros-doubles-down-on -gold/.

158 *In a DealBook column:* Steven M. Davidoff, "How to Deflate a Gold Bubble (That Might Not Even Exist)," *New York Times,* August 30, 2011, dealbook.nytimes.com/2011/08/30/how-to-deflate -a-gold-bubble-that-might-not-even-exist/.

158 *Goldline's most famous pitchman is Glenn Beck:* Eduardo Porter, "All That Glimmers," *New York Times,* November 7, 2011, http://ny times.com/2011/11/08/opinion/all-that-glimmers.html.

158 *Television business commentators predicted:* "$4,000 Gold? A Billionaire's Remarks at a Private Club Rivet New York," *New York Sun,* September 30, 2010, http://www.nysun.com/editorials/ 4000-gold/87099/; James O'Dell, "Paulson Predicts $4,000 Gold," Morgan Gold, May 5, 2011, http://www.morgangold.com/news/ 20110505-paulson-predicts-4000-gold. html.

158 *Leaving aside the holdings of gold coin speculators:* 2,250 metric tons in ETFs: Davidoff, "How to Deflate a Gold Bubble."

159 *On November 3 the Santa Monica, California, city attorney:* Stuart Pfeifer, "Goldline Metals Dealer Is Charged," *Los Angeles Times,* November 3, 2011, http://articles.latimes.com/2011/nov/03/busi ness/la-fi-goldline-charges-20111103.

159 *Goldline agreed to refund as much as $4.5 million to buyers:* Matthew Mosk, "Goldline Agrees to Refund Millions to Customers," ABC News, February 22, 2012, http://www.abcnews.go.com/ Blotter/goldline-agrees-refund-millions-customers/story?id= 15768897; Stuart Pfeifer, "Goldline Agrees to Refund up to $4.5 Million to Former Customers," *Los Angeles Times,* February 23, 2011, http://articles.latimes.com/2012/feb/23/business/la-fi -0223-goldline-settlement-20120223.

159 *As the gold price fell:* Julie Creswell, "A Gold Rush Wanes as Hedge Funds Sell," *New York Times,* September 22, 2011, http://www.ny times.com/2011/09/23/business/economy/after-huge-gains-in -gold-hedge-funds-sell.html.

159 *As it happened, this quality was crucial:* The WGC's position is laid out in three documents: "The Importance of Gold in Reserve Asset Management," June 2010; "Gold as a Source of Collateral," May 2011; "World Gold Council Position on Capital Requirements Regulation," August 2011.

163 *Barrick Gold's research department once determined:* Jim Mavor, phone interview with author.

164 *The gold price dropped like a shot crow:* London afternoon fix September 6, $1,895, London morning fix September 7, $1,844, http://www.kitco.com/londonfix/gold.londonfix11.html.

164 *Suspicion turned to Libya:* "Libyan Central Bank Sells Gold, Says Assets All Safe," Reuters, September 8, 2011, http://www.reuters.com/article/2011/09/08/libya-centralbank-idAFV9E7J300N20110908; Michael Peel, Jack Farchy, and Rouala Khalaf, "Gaddafi Regime Sold $1bn of Gold," *Financial Times,* September 8, 2011, http://www.ft.com/intl/cms/s/0/85c0912a-da1e-11e0-b199-00144fe abdc0.html#axzz2NiR9WqWu; "Gaddafi May Escape with Central Bank Gold, Says Ex-Libyan Governor," CentralBanking.com, August 25, 2011, http://www.centralbanking.com/central-banking/news/2104624/gaddafi-escape-central-bank-gold-libyan-governor.

165 *Europe was in danger of unraveling:* "In the Brussels Bunker," *Economist,* September 17, 2011, http://www.economist.com/node/215 29064.

166 *Meanwhile, American stocks lost $1.1 trillion:* Nikolaj Gammeltoft, "Dow Falls Most Since October 2008 on Economic Growth Concerns," *Bloomberg,* September 23, 2011, http://www.bloomberg .com/news/2011-09-23/dow-average-declines-most-since-october -2008-on-economic-growth-concerns.html.

166 *Bitter partisans haggled in the Congress:* See for example Jennifer Steinhauer, "Dispute on Disaster Aid Threatens Bill to Avert Government Shutdown," *New York Times,* September 21, 2011, http://www.nytimes.com/2011/09/21/us/politics/dispute-on-disaster-aid-threatens-bill-to-avert-government-shutdown.html?_r=0.

166 *In a single week it lost $200:* London morning fix September 19, 2011, $1,817; London morning fix September 26, $1,615, http://www.kitco.com/londonfix/gold.londonfix11.html.

166 *In the face of increased volatility:* Garry White, "Gold Slumps as CME Ups Margin Requirements," *Telegraph,* September 26, 2011, http://www.telegraph.co.uk/finance/financialcrisis/8788926/Gold-slumps-as-CME-ups-margin-requirements.html.

166 *The aggregate increase, according to a client note:* September 26, 2011, client note from BMO Capital Markets.

166 *Also preying on the gold price were suspicions about ETFs:* Chris Flood, "Commodities: Investor Appetite for Gold ETPs Wanes," *Financial Times,* September 23, 2011, http://www.ft.com/intl/cms/s/0/657a9358-e14d-11e0-ac59-00144feabdc0.html#axzz2NiR9WqWu; Agustino Fontevecchia, "Is GLD Really as Good as Gold?", *Forbes,* November 15, 2011, http://www.forbes.com/sites/afontevecchia/2011/11/15/is-gld-really-as-good-as-gold/; Joe Weisenthal, "MYSTERY: Why Did GLD's Published List of Gold Bars Shrink?," *Business Insider,* October 27, 2009, http://www.businessinsider.com/mystery-why-did-glds-published-list-of-gold-bars-shrink-2009-10.

168 *September 2011 closed with words like "bruising," "brutal," "turbulent":* Eric Dash and Julie Creswell, "Investor Fear over Morgan Stanley Sharpens," *New York Times,* September 30, 2011, http://www.nytimes.com/2011/10/01/business/investor-fear-over-morgan-stanley-sharpens.html.

168 *In 150 years the world supply:* From Hartwig Frimmel, head of Geodynamics and Geomaterials Research, Institute of Geography and Geology, University of Würzburg.

CHAPTER 10: SHADOW GOLD

PAGE

169 *Consider the arcane practice of fixing the price:* See note in Chapter 9 above (page 264), *Most of the trade is in the hands of a cabal of banks in London.*

171 *A group of London criminals discovered this painful truth:* Tony Thompson, "Curse of the Brink's-Mat Heist Claims Its Latest Victim," *Observer,* November 24, 2001, http://www.guardian.co.uk/uk/2001/nov/25/ukcrime.tonythompson; Mark Townsend, "Trail of Gold Grains Offers £8m Clue to Mystery of Brink's-Mat Robbery," *Observer,* June 7, 2008, http://www.guardian.co.uk/uk/2008/jun/08/ukcrime; Matt Roper, "Fool's Gold: The Curse of the Brink's-Mat," *Mirror,* May 12, 2012, http://www.mirror.co.uk/news/uk-news/the-curse-of-the-brinks-mat-gold-bullion-robbery-829220; Wensley Clarkson, "The Curse of Brink's Mat: An Ex-Cop with an Axe in His Head—And a Great Train Robber Shot Dead in Marbella," *Daily Mail,* May 5, 2012, http://www.dailymail.co.uk/news/article-2139945/Brinks-Mat-An-ex-cop-axe-head-Great-Train-Robber-shot-dead-Marbella.html; "Road Rage Murderer Kenneth Noye Loses Attempt to Appeal Sentence," BBC News, March 12, 2013, http://www.bbc.co.uk/news/uk-england-21760003.

175 *One of the sublime rituals of money:* http://www.royalmint.com/discover/uk-coins/history-of-the-trial-of-the-pyx.

175 *One such passage shook the market:* See for example Rowena Mason, "Secret Gold Swap Has Spooked the Market," *Telegraph,* July 11, 2010, http://www.telegraph.co.uk/finance/markets/7884272/Secret-gold-swap-has-spooked-the-market.html.

177 *Suspicion switched to the IMF:* Ibid.; see also for IMF as gold seller: Sandrine Rastello and Kim Kyoungwha, "IMF Sells Gold to India, First Sale in Nine Years (Update2)," *Bloomberg,* November 3, 2009, http://www.bloomberg.com/apps/news?pid=newsarchive&sid=ac4.u0JfPtWE.

177 *No sooner did opinion settle on the IMF, than the BIS itself torpedoed it:* For BIS announcement and general analysis: Carolyn Cui and Liam Pleven, "Commercial Banks Used Gold Swaps," *Wall Street Journal,* July 7, 2010, http://online.wsj.com/article/SB100014240 52748704545004575353403943560776.html.

177 *As the speculation crackled on:* For GATA conspiracy belief: "About GATA," at http://gata.org/about.

178 *In 2000 a GATA-funded litigator sued the bank:* For details of the GATA suit, see presiding judge's order posted at goldensextant. com. Scroll to "Legal Materials," click on "Gold Price Fixing Case," and select "Memorandum" and "Order for Dismissal."

178 *Chilled by the steady drizzle of apprehension:* Swap explained: Jack Farchy and Javier Blas, "BIS Gold Swaps Mystery Is Unravelled," *Financial Times,* July 29, 2010, http://www.ft.com/intl/cms/ s/0/3e659ed0-9b39-11df-baaf-00144feab49a.html#axzz2Nv 4F1Xkn.

179 *A reasonable suspicion about the gold-for-dollars swaps:* Ibid.; Cui and Pleven, "Commercial Banks Used Gold Swaps."

179 *The banks had borrowed it, mostly from their own customers:* Farchy and Blas, "BIS Gold Swaps Mystery Is Unravelled."

181 *As long as the hoard has not been compromised:* Felix Salmon, "The Problem of Fake Gold Bars," Reuters, March 25, 2012, http:// blogs.reuters.com/felix-salmon/2012/03/25/the-problem-of-fake -gold-bars/; Robert Cookson, "HK Gold Market Hit by Sophis- ticated Scam," *Financial Times,* December 2, 2010, http://www .ft.com/intl/cms/s/0/f7b05cf2-fcfc-11df-ae2d-00144feab49a .html#axzz2Nv4F1Xkn.

183 *A Zurich-based executive of Stonehage:* Martin de Sa'Pinto, "Fund View: Rich Trim Gold Holdings, Buy Art," Reuters, September 1, 2011, http://www.reuters.com/article/2011/09/01/us-stonehage-id USTRE7803NM20110901.

184 *Among those battered in the fall was John Paulson:* Julie Creswell, "Even Funds That Lagged Paid Richly," *New York Times*, March 31, 2011, http://www.nytimes.com/2011/04/01/business/01hedge. html.

184 *The* New York Times *calculated:* Azam Ahmed and Julie Creswell, "Bet on Gold Nets Paulson $5 Billion," *New York Times*, January 29, 2011, http://www.nytimes.com/2011/01/29/business/29paulson. html.

184 *His fortunes started to turn:* Kelly Bit, Christopher Donville, and Matt Walcoff, "Paulson May Deal Clients $720 Million Loss," *Bloomberg*, June 21, 2011, http://www.bloomberg.com/ news/2011-06-21/paulson-dumping-sino-forest-may-deal-clients-720-million-loss.html; Azam Ahmed, "Paulson Speaks Out on Sino-Forest," *New York Times*, June 24, 2011, dealbook.nytimes. com/2011/06/24/paulson-speaks-out-on-sino-forest/.

184 *In June that year a firm of short sellers called Muddy Waters:* Ye Xie and Victoria Stilwell, "Muddy Waters Retreats on Short Selling Chinese Stocks," *Bloomberg*, November 27, 2012, http://www. bloomberg.com/news/2012-11-27/block-gives-up-as-short-sell ing-declines-china-overnight.html.

184 *One report said Paulson was so angered:* Azam Ahmed, "Another Traumatic Month for Paulson," *New York Times*, October 10, 2011, dealbook.nytimes.com/2011/10/10/another-traumatic-month -for-paulson/.

185 *He sold a third of his company's position in the Spider:* Kimberley Williams, "Paulson & Co Holdings in 3rd Quarter: 13F Alert," *Bloomberg*, November 14, 2011, http://www.bloomberg.com/ news/2011-11-14/paulson-co-holdings-in-3rd-quarter-13f-alert .html.

185 *According to an unnamed source cited by Reuters:* Eric Onstad, "UP-DATE 3—Paulson Not Seen Deserting Gold After $2 Bln ETF

Sale," Reuters, November 15, 2011, http://www.reuters.com/article/2011/11/15/gold-paulson-idUSL5E7MF1SS20111115.

185 *The Oxford dictionary: The Shorter Oxford English Dictionary* (Oxford: Oxford University Press, 1973).

185 *Gold is a doubtful sanctuary:* Jan Harvey and Amanda Cooper, "Gold Tumbles as EU Leaders Reassure on Greece," Reuters, September 15, 2011, http://www.reuters.com/article/2011/09/15/markets-precious-idUSL5E7KF1A420110915; Jan Harvey, "Gold Turns Lower as Wider Markets Slide," Reuters, October 4, 2011, http://www.reuters.com/article/2011/10/04/markets-precious-idUSL5E7L419F20111004.

185 *In* The Golden Constant: Roy Jastram, *The Golden Constant: The English and American Experience, 1560–1976* (New York: John Wiley & Sons, 1977), 177.

CHAPTER 11: THE GOLD IN THE BAMBOO FOREST

PAGE

189 *He sits in audience:* Al-Bakri, trans. J. F. P. Hopkins, in *Corpus of Early Arabic Sources for West African History*, eds. J. F. P. Hopkins and N. Levitzon (Cambridge: Cambridge University Press, 1981), 80. Bida legend: http://www.bbc.co.uk/worldservice/africa/features/storyofafrica/4chapter1.shtml.

189 *In 1698 the Dutch traveler William Bosman:* David Crownover, "An Ashanti Soul-Washer Badge," *Expedition* 6, no. 2 (Winter 1964): 11–12, http://www.penn.museum/documents/publications/expedition/PDFs/6-2/An%20Ashanti.pdf.

190 *A Danish doctor wrote:* T. Edward Bowdich, *Mission from Cape Coast Castle to Ashantee* (London: John Murray, 1819), 1.

190 *The ground of our resting place was very damp:* Ibid., 20.

191 *They marched in at two o'clock in the afternoon:* Entry to Kumasi, festivities, meeting the Asantehene: Ibid., 31–41.

193 *The king's soul washers wore gold disks:* Crownover, "An Ashanti Soul-Washer Badge." See also "A Soul Washer with Gold Disk and Special Helmet," http://www.aluka.org/action/showMetadata?doi=10.55 55/AL.CH.DOCUMENT.BFACP1B10189.

193 *The linguists were their living archive:* See for example http://www .metmuseum.org/toah/works-of-art/1986.475a–c. The antler finial is in the Gold of Africa Museum, Cape Town.

194 *Where is the Gold Stool?:* "The Story of Africa: West African Kingdoms," BBC, http://www.bbc.co.uk/worldservice/africa/features/ storyofafrica/4chapter6.shtml.

195 *But that night, in a secret meeting:* Yaa Asantewaa: *The Oxford Encyclopedia of Women in World History,* 2008, http://www.oxfordrefer ence.com.

199 *The Mali Empire was unknown to Europeans until the appearance of its greatest ruler, Mansa Musa:* See for example Hopkins and Levitzon, *Corpus of Early Arabic Sources,* 341–47, and A. J. H. Goodwin, "The Medieval Empire of Ghana," *South African Archaeological Bulletin* 12, 1957: 108–12.

201 *Mark Nathanson was the son of a wholesale grocer:* I relied largely on Larry Phillips for my account of Nathanson, with a few details from Jean Kaisin. See also Kenneth Gooding, "Sadiola–Nathanson's luckiest strike," *Financial Times,* February 9, 1996. The Nathanson Centre has a biographical note at http://nathanson.osgoode.yorku .ca/about/founding-benefactor/; http://www.iamgold.com/english/ corporate/iamgold-milestones/default.aspx.

201 *Treasure hunters have searched for a fabulous city called Ophir for thousands of years:* Marx, *The Magic of Gold,* 18.

CHAPTER 12: KIBALI

216 *The Lord's Resistance Army:* "Hundreds More Flee Continuing LRA Attacks in North-East Congo," UNHCR, March 30, 2012, http://www.unhcr.org/4f75a5589.html. This account demonstrates that one year after our visit the killing was still going on.

217 *Australian prospectors had discovered the gold in 1903:* See for example Brandon Prosansky, "Mining Gold in a Conflict Zone: The Context, Ramifications, and Lessons of AngloGold Ashanti's Activities in the Democratic Republic of the Congo," *Northwestern Journal of International Human Rights* (Northwestern University School of Law) 5, no. 2 (Spring 2007): 236–74, http://www.law.northwestern.edu/journals/jihr/v5/n2/4/Prosansky.pdf.

217 *A bloodbath followed:* Ian Fisher and Norimitsu Onishi, "Chaos in Congo: A Primer. Many Armies Ravage Rich Land in the 'First World War' of Africa," *New York Times,* February 6, 2000, http://www.nytimes.com/2000/02/06/world/chaos-congo-primer-many-armies-ravage-rich-land-first-world-war-africa.html?pagewanted=all&src=pm.

217 *In August 1998 Ugandan soldiers occupied the goldfield:* Anneke van Woudenberg, *The Curse of Gold,* Human Rights Watch, 2005, 15–18, http://www.hrw.org/reports/2005/drc0505/drc0505.pdf.

218 *Then their captors roped them together:* Christopher Rhoads, "Peacekeepers at War," *Wall Street Journal,* June 23, 2012, http://online.wsj.com/article/SB10001424052702303836404577476460542151978.html.

218 *Similarly, failed pumps meant that miners had to wade for miles:* Van Woudenberg, *The Curse of Gold,* 54.

218 *The Ugandans took $9 million worth of gold:* Ibid., 15.

218 *Their successors did even better:* Ibid., 55.

219 *Sir Sam Jonah was a celebrated gold miner:* See for example "Sam Jonah: The Man with the Golden Touch," *Sun* (Lagos, Nigeria), April 13, 2013, http://sunnewsonline.com/new/specials/aspire/sam-jonah -the-man-with-the-golden-touch/; executive profile at *Bloomberg Businessweek,* http://investing.businessweek.com/research/stocks/ private/person.asp?personId=372443&privcapId=35042515; Simon Robinson, "Sam Jonah: AngloGold Ashanti," *Time,* December 17, 2004, http://www.time.com/time/magazine/article/0,9171,100 9745,00.html; http://www.anglogold.com/Additional/Press/Ashanti /2003/Ashanti+Chief+Executive+Sam+Jonah+Knighted.htm; "Sam Jonah Moves to Non-Executive Board Position at Anglo-Gold Ashanti," http://www.anglogold.com/additional/press/2005/ sam+jonah+moves+to+non-executive+board+position+at+anglo gold+ashanti.htm.

219 *He had taken Ashanti into Kilo-Moto nine years earlier:* Van Woudenberg, *The Curse of Gold,* 60.

220 *AngloGold funded one such trip:* Ibid., 61.

220 *The company said it had only paid $8,000 out of petty cash:* "Anglo-Gold Ashanti's Activities in the Democratic Republic of the Congo," statement from company, 6, http://www.anglogoldashanti.co.za/ NR/rdonlyres/ECBCA20E-5B45-4625-B4E7-B8517F369043/0/ AGA_and_the_DRC.pdf.

220 *But Toronto is the world's leading gold-mine city:* http://www.tmx .com/en/listings/sector_profiles/mining.html.

220 *Regardless of its size, Moto was the cat's paw of important interests:* Barry Sergeant, "Inside Moto Gold Mines," *Mineweb,* April 1, 2007, http://www.mineweb.com/mineweb/content/en/mineweb-histori cal-daily-news?oid=18928&sn=Detail.

222 *With its joint venture partner, AngloGold Ashanti, Randgold took a run at Moto:* See for example William MacNamara, "Randgold in $300m Placing for Expansion," *Financial Times,* July 29, 2009,

http://www.ft.com/cms/s/0/e4a66744-7bd6-11de-9772-00144fe
abdc0.html#axzz2OTZpnjSf; Susan Kirwin, "Randgold Outbids
Red Back for Moto Goldmines," *Northern Miner,* July 27, 2009,
http://www.northernminer.com/issuesV2/VerifyLogin.aspx.

223 *Bristow worked on his counterbid:* All from Mark Bristow and Grant
Bristow, interviews with author.

223 *Red Back's Lukas Lundin:* Eduard Gismatullin, "Billionaire Swedes
See Lundin Jump After Matching Apple: Energy," *Bloomberg,*
May 28, 2013, http://www.bloomberg.com/news/2013-05-27/
billionaire-swedes-see-lundin-jump-after-matching-apple-energy
.html.

224 *He had sold the Mauritanian property:* "Canada's Kinross to Buy
Red Back for $7.1 Billion," *New York Times,* August 2, 2010,
http://dealbook.nytimes.com/2010/08/02/canadas-kinross
-to-buy-red-back-for-7-1-billion/.

224 *Tye Burt:* Interview with author; for figures on possible reserves of
Tasiast, and advancement of Kinross in rank of producers, see for
example Stephen Grocer, "Deal Profile: Kinross to Acquire Red Back
Mining," *Wall Street Journal,* August 3, 2010, http://blogs.wsj
.com/deals/2010/08/03/deal-profile-kinross-to-acquire-red-back
-mining/.

224 *Kinross fired Burt:* Liezel Hill, "Kinross Gold Fires CEO Tye Burt,
Replaces with Rollinson," *Bloomberg,* August 2, 2012, http://
www.bloomberg.com/news/2012-08-01/kinross-gold-fires-ceo-tye
-burt-replacing-him-with-rollinson.html; Liezel Hill, "Kinross Gold
Takes $3.1 Billion Writedown on Tasiast Mine," *Bloomberg,* Feb-
ruary 14, 2013, http://www.bloomberg.com/news/2013-02-13/
kinross-gold-takes-3-1-billion-writedown-on-tasiast-mine.html.

225 *Its soldiers come from countries:* Rhoads, "Peacekeepers at War."

225 *Because the United States' Dodd-Frank Act:* Edward Wyatt, "Use of
'Conflict Minerals' Gets More Scrutiny from U.S.," *New York Times,*

March 19, 2012, http://www.nytimes.com/2012/03/20/business/use-of-conflict-minerals-gets-more-scrutiny.html?pagewanted=all.

226 *A human rights group calculated:* Gouby, "Gold Now Top Conflict Mineral in Congo," Associated Press, October 25, 2012.

226 *In eastern Congo, a war that has already killed 3 million people:* Jeffrey Gettelman, "Troops Mass in Fought-Over City, Raising Fear of New Violence in Congo," *New York Times,* December 16, 2012, http://www.nytimes.com/2012/12/17/world/africa/troops-mass-in-fought-over-city-raising-fear-of-new-violence-in-congo.html; 3 million dead: van Woudenberg, 12.

226 *Randgold priced its reserves:* http://www.randgoldresources.com/randgold/content/en/randgold-kibali-project. See second footnote to table.

227 *It's not clear that our distant ancestors:* Interview with Ben Roberts, at the time, curator of the European Bronze Age collections at the British Museum, London, and now on the faculty of Durham University.

228 *One theory about the super-rich:* Chrystia Freeland, "The Self-Destruction of the 1 Percent," *New York Times,* October 13, 2012, http://www.nytimes.com/2012/10/14/opinion/sunday/the-self-destruction-of-the-1-percent.html?pagewanted=all.

INDEX

Matthew Hart is a veteran writer and journalist, and author of seven books, including the award-winning *Diamond*. His work has appeared in *The Atlantic Monthly*, *Granta*, *The Times* and *The Financial Post Magazine*. He was a contributing editor of the New York trade magazine *Rapaport Diamond Report* and has appeared on *60 Minutes*, CNN, and the National Geographic Channel. He lives in New York City.